The Lit Environment

Other books by the same author

Lighting modern buildings
Lighting historic buildings

The Lit Environment

Derek Phillips

With a foreword by

Professor Max Fordham

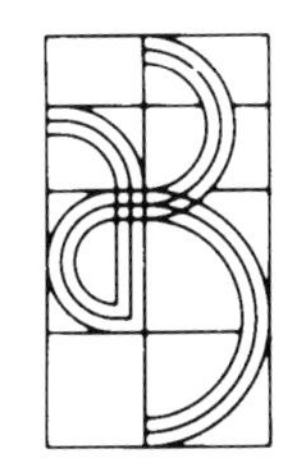

Architectural Press

OXFORD AUCKLAND BOSTON JOHANNESBURG MELBOURNE NEW DELHI

Architectural Press
An imprint of Butterworth-Heinemann
Linacre House, Jordan Hill, Oxford OX2 8DP
225 Wildwood Avenue, Woburn, MA 01801-2041
A division of Reed Educational and Professional Publishing Ltd

A member of the Reed Elsevier plc group

First published 2002

British Library Cataloguing in Publication Data
Phillips, Derek, 1923–
The lit environment
1. Lighting, Architectural and decorative
2. Lighting, Architectural and decorative – Case Studies
I. Title
729.2′8

Library of Congress Cataloguing in Publication Data
Phillips, Derek
The lit environment/Derek Phillips; with a foreword by Max Fordham
p. cm.
Includes bibliographical references and index
ISBN 0-7506-4889-9
1. Lighting, Architectural and decorative. 2. Street-lighting.
3. Municipal lighting. I. Title.

TK 4188 .P49 2001
621.32–dc21 2001053897

ISBN 0 7506 4889 9

Composition by Genesis Typesetting, Laser Quay, Rochester, Kent
Printed and bound in Great Britain by Bath Press – Glasgow

Contents

Acknowledgements

Since the inspiration for this book was the first lecture in the series devoted to the memory of Jack Waldram, it is appropriate that I should acknowledge my debt to him, as arguably the first lighting engineer in this country to be recognized as a lighting designer, understanding the needs of architects and of architectural space.

It would be wrong not to thank the many architects who have provided information on their projects, but too long a list; let them be known by their lit environments. To these must be added the many lighting designers and manufacturers, who have been of assistance.

It is important to thank all those who have commented on the various chapters of the book; including David Loe, Professor Malcolm Parry, Carl Gardner, Bill Crawforth and Bob Divall, but most of all my old colleague John Howard, who has, like me, retired and who generously held my hand over the chapter on Tactics; and finally to thank Marie Armstrong who was responsible for turning my inadequate sketches into understandable line drawings.

Half of the photographs in the book are derived from my own slide library built up over forty years in practice as an architect and lighting designer, a slide archive supplemented by friends and constantly updated. My thanks go to all who have contributed.

Foreword

This book is one of a set about lighting design. It addresses the issues raised by the perception of light experienced by people who are outside buildings. It is not simply about floodlighting buildings but takes a more embracing approach to deal with light for the whole external environment.

Natural and artificial light are both effective by night and by day. A night scene bathed in a full moon represents natural light at night while sign lighting is artificial light used in a bright daylit environment.

Buildings during the day are generally lit with diffuse light, whereas at night they can be presented in a new way using strongly directed light. Any external environment extends upwards to the whole universe (except when the sky is obscured by clouds). It is also limited by the ground surface and the outside of buildings.

The streets extend away from a square and are a part of the whole problem of lighting a square. The need for consistency in built form has led to planning legislation, but a similar consistency in the lit environment is difficult to achieve because of the current anarchy which allows each built artefact to be lit without respect for its neighbours.

A book on lighting represents a strong fusion of science and quantity with light and vision. Visually inclined designers should understand how science can improve their vision and engineers with a strong science basis should understand how they must come to terms with visual affects.

Ever since artificial lighting was invented, its design has attracted individuals who do not take an extreme engineering/architecture position. The number of past presidents of the Institute of Light and Lighting who are also Fellows of the Royal Society shows that science has always been a part of lighting design. At the same time many lighting engineers are also architects and Derek Phillips is one.

Buildings, and hence architecture, are the outstanding artifacts of any cultural system. The design of modern artifacts has to achieve a series of complicated functions ranging from providing a pleasurable emotional response for the user to meeting a whole set of functional requirements that users expect to take for granted. This range of functions is leading to a tendency to develop specialists. There are those who develop a purely subjective set of skills and imagine wrongly that the physical realities have a low priority, and there are those who concentrate on the physical behaviour and ignore the subjective responses. At the extremes this specialization is stultifying and will prevent the design process from meeting the challenges of the twenty-first century. To meet the challenge of climate change, it is clear that the conceptual design of buildings must be modified. Electricity for lighting is often a main part of the demand that leads to the production of carbon emission in our society. This means that lighting is one of the most important aspects of designing the built environment of the future.

Professor Max Fordham
OBE, MA(Cantab) FREng, FCIBSE, FconsE Hon, FRIBA

Preface

The first in the series of Waldram Lectures in 1990,[1] titled *City Lights* attempted to set out a framework for the nighttime lighting of cities:

> . . . to instill an overall concept which can be applied to the visual appearance of the city as a whole, something to work towards. To inspire the imagination of those who can influence such matters, to work towards making our cities places of wonder, places with a variety of atmosphere, appropriate to the infinite variety of city life; then perhaps the Greek vision of the city being the highest manifestation of civilisation could become a reality.

In an excellent little booklet by the Fine Arts Commission[2] Trafalgar Square was described as 'more nightmare than nightscape' and this sadly was neither alone nor the worst example. It is sad to think that London, one of the greatest capital cities in the world, has some of the most unappealing lighting; whereas a provincial city in France, such as Lyon, comes alive at night with beautifully lit squares and buildings.

The purpose of this book, the third in a series of publications on Lighting by Architectural Press, is designed to flesh out the theories set out in the Waldram Lecture and to offer a general theory on lighting at night, not just for the city, but equally for the town and countryside.

It would be wrong to paint a completely dark picture, as there are isolated instances where good work has been done associated in most instances with the Millennium – such as the refurbishment of Trafalgar Square, the regeneration of the harbour area of Bristol or by the Church Floodlighting Trust; but there is still a lack of cohesion, no overall concept. If this book does something to make things work towards a general improvement of the lit environment in our towns and countryside it will have served its purpose.

Derek Phillips

---

[1]Phillips, D. *The Waldram Lecture, London* (1990) Sponsored by CIBSE and ILE.
[2]A Report by the Royal Fine Art Commission (1994) *Lighten Our Darkness*.

Introduction

The exterior lighting brief . . . identifying the elements . . . analysis of building form . . . 'a little light goes a long way at night' . . . the lit environment . . . lighting equipment

THE BRIEF

It will be unusual for an architect and his lighting designer to be asked to design the lighting of a whole 'lit environment'; except in some large-scale developments. The lit environment is composed of so many elements: roads, buildings, incidents such as statues and fountains, and soft landscape. Each element may have been considered separately; but it is only when these are coordinated into a whole that a satisfying night environment or 'nightscape' is likely to be achieved.

When establishing the lighting designers brief, it is necessary to identify the distinguishing features of the environment experienced during the day, to ensure that these are understood. This is not to suggest that the nightscape must necessarily attempt to duplicate the daytime experience, far from it; the appearance at night will have its own aesthetic, derived from the many possibilities which modern lighting techniques permit. There may well be a case for playing God by diminishing the effect of certain less satisfactory elements experienced during the day, whilst emphasizing others; but at the end of the day it should appear to be the same place.

It is important that the concept of floodlighting is put in perspective. For many traditional buildings, light directed on to its surfaces by means of well-directed and concealed light sources will be a suitable solution, but this is far from universal; as modern buildings are developed with more glass than solid wall, light directed from outside may no longer be feasible.

Traditional buildings designed for their internal function, and as experienced by daylight, were generally more solid than void, but now the building at night may take on an appearance which has little to do with external light sources.

There are a number of criteria that must be considered in the exterior lighting brief, apart from aspects of safety and security, which go without saying. The first must be the nature of the site and its surroundings, whether city, town, village or countryside; its social implications, whether

DP Archive

Manufacturer's Trust Building, 5th Avenue, New York, by day (1952)

DP Archive

Manufacturer's Trust Building, 5th Avenue, New York, by night (1952)

quiet residential or city centre. The lighting of an isolated country church may be accomplished with very low light levels there being little competition from either street lighting or other buildings . . . but the old cliché that 'a little light goes a long way at night' should not be forgotten. On the other hand, the situation is quite different in the city centre, surrounded by theatres and leisure buildings, where high levels of lighting will be dictated by the different interests vying for attention.

The nightscape of an environment will be informed for the most part by the appearance of the surrounding buildings, and it has only been in the twentieth century that the possibilities of artificial lighting have given expression to the 'Second Aspect of Architecture', or how the building is experienced after dark.

The function of a building will to a large extent help to establish its night appearance. This itself may differ from early evening to late at night; for example during working hours an office building will clearly need to establish itself in its location, while once the office is closed or is only partly in use, the 'advertising' character of the lighting will be diminished, and considerations of energy conservation should dictate a lower profile.

On the other hand, leisure buildings may wish to advertise their presence long into the night whilst residential areas require only to satisfy the needs of safety and security . . . different emphasis at different times.

Whilst each building will require individual treatment, there are some aspects which need to be considered by the lighting designer in all cases . . . the modelling of the exterior, the colour of the light sources, the location of the equipment and the danger of glare and light pollution. The principles associated with the needs of different forms of building are dealt with under the following headings: Solid form; Glazed form; Mixed form – solid and void; and Transparent form.

Although in most cases buildings will be seen as the most important elements of the lit environment in towns, it will consist of others comprising the 'Spaces between' . . . the roads, pathways, pedestrian areas, parks, landscape, transport, and areas associated with water or sport. Then there are the 'incidents', such as walls and gateways, bridges, fountains and statues, together with the soft landscape, of trees, grass and parkland, which may be seen as lighting opportunities, adding delight and excitement to nightscape. Finally, the 'ephemera' of lighting which can add sparkle and excitement to our towns, in which are included the fireworks and traditional celebrations of the seasons.

The lighting of roads has been thought of in engineering terms, and whilst there is a clear necessity to design the lighting of roads to meet the needs of safety and identity, the systems employed often lack cohesion. There is lack of an overall consistency of approach. On descending into London's Heathrow airport at night the view below presents a disorganized picture of London, which cannot be explained entirely by the lack of cohesive planning, although a similar view of a new town such as Milton Keynes, exhibits simplicity and coordination.

When Colin Buchanen gave his famous paper *Traffic in Towns* in 1963 he identified the problems, arguing for a coherent strategy leading to a network of road systems classified as being of varying importance; here lighting design could have played a crucial role, but the opportunity was missed. In 1990 it was said that the only viable solution to the problem of traffic in London was 'a vastly improved public transport system . . . and limited car access' since then the problem has deteriorated still further (Waldram Lecture, London, 1990).

Having stressed the necessity to provide a coherent traffic system for cars, it is important to plan separate routes for pedestrians, and also if possible for bicyclists, where the levels of lighting can be relatively low. Lighting can help identify pedestrian routes, and to link these up with other pedestrian open spaces, so that walking around towns becomes a safe and pleasurable experience. Some of the Dutch towns display an excellence, where the canal system by its nature is separated from the road traffic, and the pedestrians have their own routes.

Open spaces vary from the great piazzas, such as St Mark's, Venice, to the small left-over spaces amongst residential developments, such as the London squares or space for the town market, which comes alive only once or twice a week. The approach to the lighting of such areas should reflect their use, where the lighting of the surrounding buildings plays an important role.

Bridges can provide a symbolic emphasis, particularly when associated with water, due to the multiple reflections which vary with the time and the weather . . . these vary from the huge road or railway bridge to the small pedestrian bridge and there is a need to establish the visual form of the bridge as an important element of nightscape.

Much of the delight in the nightscape of our towns derives from the incidents, such as statues or fountains, which create lighting opportunities; to close a vista or make an emphatic statement – when these are disregarded they disappear at night, an opportunity missed. Small towns in Europe benefit enormously from such minor incidents. Forgetting the great fountains and statues of the past, much can be accomplished with minimal means to enliven and enrich our towns. The weather and temperature in France does not greatly differ from the British Isles, yet French towns seem to have little difficulty in keeping fountains running throughout the year.

This is not to forget the 'ephemera' of lighting for special occasions, fireworks and Christmas lighting – all contributing to the excitement and magic of town life. From earliest times there was a functional need to emphasize gateways in towns, and the tradition has continued, often associated with enclosing walls. Such incidents helped to establish the character of a town in the days when they were protected by surrounding walls for security.

The parks and open spaces in towns and cities provide welcome interludes between the roads and buildings; and those towns which lack such facilities are the poorer for it. London, for example, has the royal parks to provide the 'lungs' of the city to find a home for trees, grass, planting, and wildlife, and were it not for the fact that these are royal parks they would undoubtedly have surrendered to the bulldozer by now. Parks where people can wander after dark, and enjoy the open space, add immeasurably to the quality of town life, and this is another area where safety and security are important, but where lighting can provide more than this.

The spaces between are the key to the whole strategic planning of the lighting of towns and cities. Driving or walking through a town should be a sequential visual experience, varying with time and season, aimed at providing delight, enchantment and excitement. Lucky the lighting designer who has the possibility of associating the lighting of buildings, bridges or landscape with water. There is a charm about light and water which together can enhance any scene . . . indeed, it must be difficult not to succeed.

LIT EXTERIOR ENVIRONMENT

A good example of the lit environment is the Tivoli gardens in Copenhagen with its ornamental lakes, restaurants and concert hall which were created over 50 years ago. The lighting design is constantly updated, using the latest lighting technology to achieve a magical effect, whilst always working within a recognizable design theme ensuring that the 'unity' remains. The design theme is essentially traditional, and has a classical appeal to the many generations who enjoy the atmosphere. It has always been a surprise to the author that the concept of Tivoli has not been matched elsewhere; the climate in Denmark is no more suitable for such nighttime activity throughout the year than in London (see Chapter 6, Lit environment, p. 111).

While it would clearly be possible to analyse the nighttime appearance of a parkland area such as Tivoli in terms of the individual elements of buildings, lakes, fountains, and walkways, where these have been planned together there is an additional factor – the coordination of building materials and form, colour, modelling and scale to create a unity of experience . . . a lit environment.

While a single shop would be considered an individual element in a shopping street – where there are rows of shops, footpaths, roads, car parking and lighting systems – the whole would present a lit environment. This is the case with some of our new shopping precincts (designed as a whole they can be successful) whilst one only has to visit many of the shopping streets in our older towns to see the unfortunate results where there are different and conflicting interests involved. The results are only marginally better in central business areas, where one might have expected some cohesion and unity to exist.

No one lit environment will be identical to another, but it is possible to identify a number of generic types as follows:

- City centre project
- Educational establishment, University campus
- Shops and shopping complex
- Hotels and leisure complex
- Marine, riverside and sea frontage
- Industrial
- Residential areas

The successful lit environment is the epitome of nightscape where everything has come together; and where this has been achieved, it is to the great credit of the architects, lighting designers and public authorities who have overcome the difficulties of coordination, economics and inertia to achieve their goal.

LIGHTING EQUIPMENT

The type and location of light sources is at the nub of nightscape; and this, coupled as it is with problems of glare and light pollution, are clearly some of the most difficult issues for the lighting designer to resolve. It is easy to solve the floodlighting design for a church on an isolated site by locating the light sources around the outside of the church, if one ignores the appearance of the installation during the day. Many of our churches appear satisfactory at night, but fail the daytime appearance test due to obtrusive lighting equipment set at ground level. The fittings should either be recessed into the ground, or shielded from normal views above ground.

Modern light sources are capable of providing an almost infinite variety of colour and whilst this can be an advantage in the right hands and used sparingly; in the wrong hands it can produce the most bizarre effects.

It is not intended to go too deeply into the many forms of external lighting equipment available to the designer, other than to emphasize that they exist; light sources are in a constant state of development, together with their associated hardware.

However, the purposes for which the lighting equipment is needed do not change greatly, so discussion needs to be concentrated on the 'why' rather than the 'how' of lighting the exterior environment.

Part 1

1 Strategy

History of exterior lighting ... problem analysis ... buildings ... roads and pathways ... spaces between ... incidents ... light pollution ... visual masterplan

HISTORY

The nighttime lighting of our towns and cities is not only about street lighting, very far from it, but perhaps it is fair to say that this is where it all started.

Streets as far back as ancient Rome were dangerous places at night, and the first 'street lighting' took the form of 'flares' carried by those wishing to venture abroad at night – a system lasting well into the seventeenth century.

As early as 1415 in England it was decreed that anyone owning a property rated at more than £10 per year, was obliged to burn candles outside their house to provide a little illumination for security purposes; but street lighting, as we understand it, started some 300 years later at the end of the seventeenth century when oil lamps were placed on wooden poles; being replaced in the early eighteenth century by cast iron brackets.

The City of Birmingham was very advanced and in 1711 ordered the installation of 700 street lights, but these would still have been powered by traditional lamps using animal or fish oil.

It was not until the beginning of the nineteenth century, in 1807, that the engineer Albert Winsor put up his famous gas lamps in Pall Mall on thirteen elegant hollow iron lamp posts. By the twentieth century street lighting had become well-established; with much ingenuity being directed towards the design of the poles and brackets on which the lamps were mounted, with the gas source subsequently being replaced by electric lamps.

DP Archive

Decorative street lighting 'multiple bracket': Central Square, Brugge

PROBLEM ANALYSIS

The strategy for exterior lighting is complex, and it helps to break this down into a series of separate subjects, for whilst all are interrelated in the sense that their solutions should add up to a unified whole, each in its own way is the subject of a separate discipline.

The subjects to consider are as follows:

The Buildings, the Roads and Pathways, the Spaces between and Incidents

The objectives of an exterior lighting strategy are as follows:

1. To provide a safe and secure environment for people
2. To create safe routes for traffic, cyclists and pedestrians
3. To facilitate the extended use of parks, open areas and sports facilities
4. To enhance significant elements and key points . . . to create lighting opportunities
5. Finally, by means of a 'Visual Masterplan' to ensure the coordination of all the elements in unity.

Decorative street lighting 'multiple bracket': typical fitting in Malta

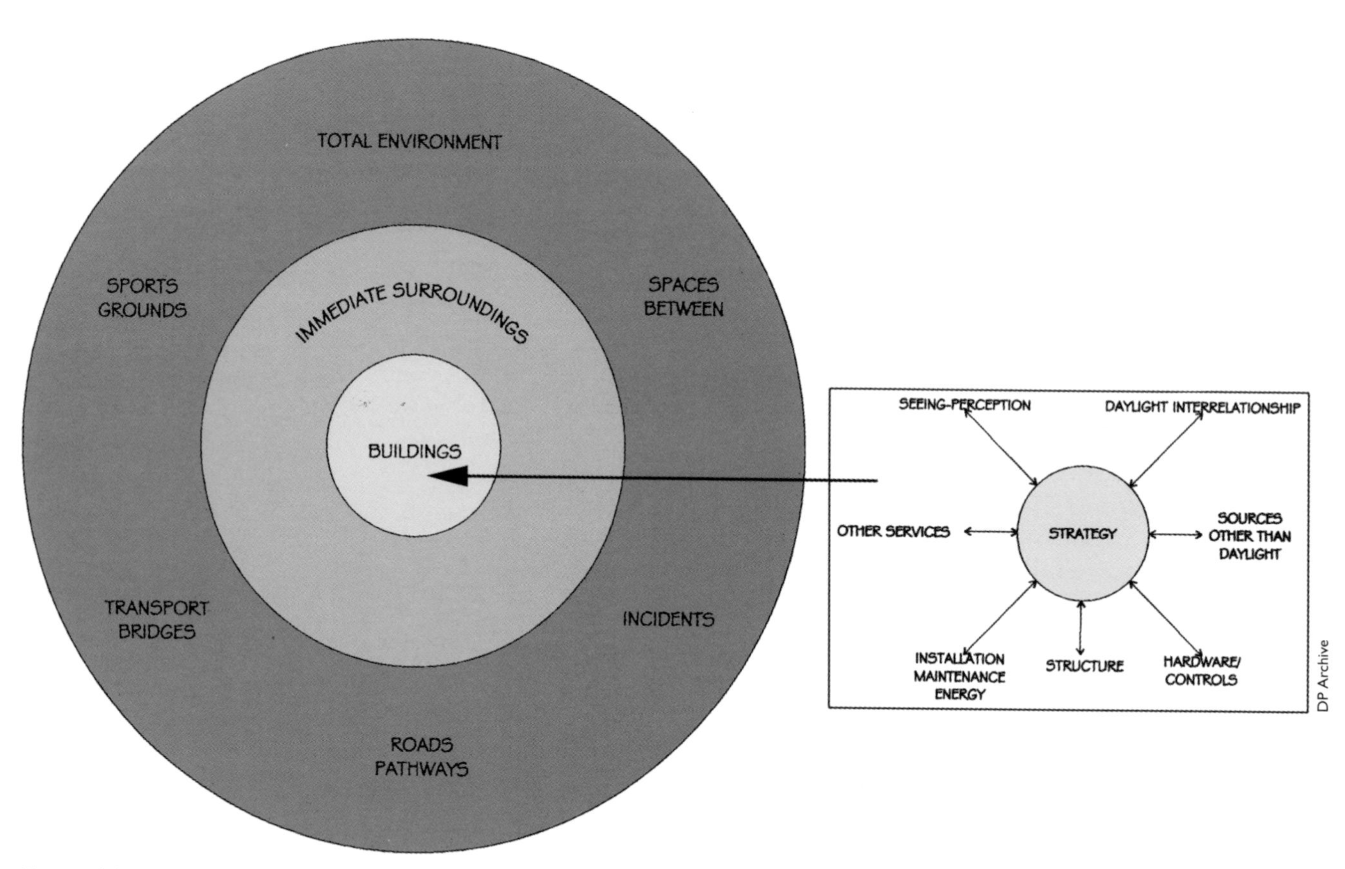

Exterior lighting strategy

BUILDINGS

Increasingly, since the beginning of the twentieth century, 'nightscape' has been used to describe the lighting of buildings, streets, open spaces and incidents, which form the overall infrastructure of the night environment.

Initially, the form of buildings after dark would have been perceived by the light from their windows, although often heavily curtained, with other light perhaps providing accent at entrances and gateways. The buildings themselves were designed to permit the entry of daylight during the day, with window openings punctured in solid walls.

The view during the day would have been of solid structure; whilst at night the building itself would have virtually disappeared, relying perhaps on the availability of moonlight to express its identity.

However, for various reasons – from a sense of security, perceived social significance or perhaps due to a sense of history – some buildings were chosen for emphasis at night; at first the only method available was external floodlighting.

In the early days, floodlighting by employing a gas source and later electricity, had a momentum of its own which tended to obscure any logical criticism of the results; results which by today's standards were often far from satisfactory. The initial solution was to throw as much light as economically feasible on to the façade of a building, irrespective of its form; this tended to lead to the building appearing as a cardboard cut-out, lacking modelling and diminishing its architectural integrity.

When it was recognized that this approach was aesthetically unsatisfactory, attempts were made to provide depth of modelling of the exterior by locating light sources in such a manner as to create desirable shadowing of the façade; but the danger here lay in unacceptable glare from the light sources themselves. However, despite the criticism, the lighting of buildings at night has emerged as an important technology, and one where, by careful engineering, the results have become satisfactory.

The lighting of buildings at night, even in an energy-saving ethos, has expanded to encompass greater areas of our towns and cities.

In the late 1950s the Philips International Lighting Review (ILR) published a seminal article in which the Italian architect Gio Ponti articulated the idea of the 'Second Aspect of Architecture', illustrating this by his designs for the Pirelli building in Milan.

If the first aspect was the appearance of a building during the day, as revealed by natural light with all its variations due to season, weather and cloud conditions, then the second aspect is its appearance by night; this calls for an integration of the architectural planning of a building with new forms of structure, recognizing, as it were for the first time, that the night appearance of a building is of equal significance, and should be considered as carefully as its daytime counterpart; this concept has had a significant influence. Up to this time it would have been more likely that a building would have been lit at night by some form of traditional exterior floodlighting.

This type of lighting was certainly appropriate for traditional buildings where the glazed area of the façades was small compared with the solid areas of masonry. The development of modern structural methods on the other hand now permitted very large areas of glazing or voids, so that the architect was able to express the design of the building in an entirely new way.

(a)

(c)

(b)

The Pirelli building, Milan, at night (a) Front view; (b) back view; (c) side view (Architect Gio Ponti)

DP Archive

Façades of buildings might now be composed entirely of glass voids; although for a number of reasons, not least, those of environmental control, the likelihood is for a balance to be struck with areas of solid and areas of glazing, or a 'mixed form'. This has had its effect upon the nighttime appearance of buildings and this will be explored further in Chapter 2 (Buildings).

The enormous developments in modern lighting technology now make it possible for almost any appearance of a building at night to be designed and engineered . . . higher or lower levels of light can be related to the setting, whether it is for a building in a brightly lit town, or an isolated church or Public House in a small village.

The colours of light sources now provide a wide palette to ensure compatability of light with building materials, so that the warmth of the light source can be related to the warmth of old stone or brickwork; while many modern structures of concrete may be more appropriately lit with cooler sources.

Courtesy of International Lighting Review

(a)

Courtesy of International Lighting Review

(b)

The Guggenheim Museum, New York. Comparison of (a) day and (b) night views (Architect Frank Lloyd Wright)

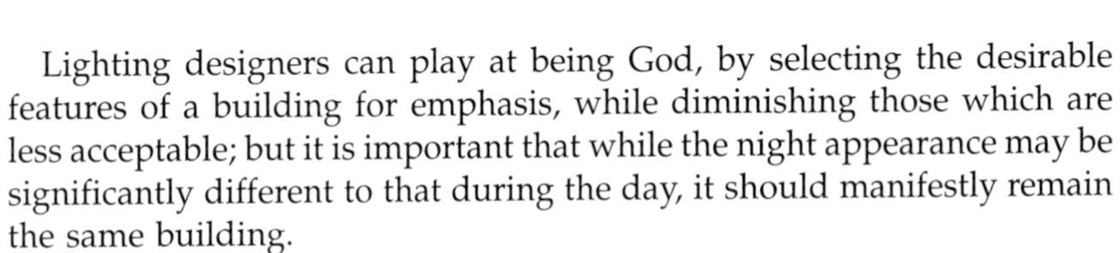

Lighting designers can play at being God, by selecting the desirable features of a building for emphasis, while diminishing those which are less acceptable; but it is important that while the night appearance may be significantly different to that during the day, it should manifestly remain the same building.

It is clear that in the right hands the opportunities for successful design are constantly increasing; but therein lies the root of the problem; there is an equal probability, if left in the wrong hands, for a design to go horribly wrong. The form of a building can be destroyed, bizarre colour can be employed, and what in daylight appears as an entirely satisfactory building, can at night be turned into a nightmare.

ROADS AND PATHWAYS

The criteria for the design of our roads and pathways are as follows:

1. To assist the provision of a coherent traffic system for cars, where orientation is assisted by visual emphasis.
2. To plan routes for pedestrians, related to adequate public transport systems, in a secure environment.
3. To plan possible cycleways free from the dangers of car traffic.

These requirements are the basic engineering considerations, and too much attention has, in the past, been paid to the engineering function, and insufficient to the creation of a satisfactory visual environment – at night the realm of the lighting designer amongst others.

There is a fundamental difference between the lighting required for cars and that for pedestrians; as the great Jack Waldram (in a paper to the Illuminating Engineering Society) observed . . . 'the pedestrian is a slow-moving highly manoeuvrable subject, for whom the lighting intensity may be relatively low . . . whilst the motor car is a fast and unmanoeuvrable object not inclined to linger, requiring significantly different conditions', he might have added . . . 'carrying its own light on its back'.

DP Archive

Badly lit building of a library, London by day

DP Archive

Badly lit building of a library, London by night

In an ideal situation, the two would be kept separate, and in our new towns much has and can be done to overcome the problems of conflicting priorities; but it is in our older towns that the worst problems exist.

It is clearly not possible for the problems of planners to be solved entirely by lighting design; but where those responsible for planning our roads are a part of the same team, there are things which can and should be done.

The lighting problems can be considered under the following areas: glare, orientation and speed.

Glare

As with the lighting of buildings, or one might say with any lighting solution, glare is the principal menace. The problem with street lighting is associated with the height and spacing of lanterns.

Street lights may be placed high and far apart, or lower and closer together; the lanterns being designed to spread the light to provide a given level of light in between. In an attempt to achieve more economic solutions the lanterns may be placed too far apart in an attempt to distribute the light further between them, causing a classic glare situation. Glare raises a person's adaptation level making it more difficult to provide acceptable visual conditions.

This may be coupled with equipment design, when a misunderstanding of the distribution characteristics of a fitting can lead to the spillage of light where it is not wanted; while the same piece of equipment would be perfectly satisfactory if used in the right situation.

On the continent there is a greater awareness of the visual aspects of the street scene where 'cut-off' lanterns are used. These, while reducing the level of light between the lamps' diminish glare, providing more comfortable conditions.

Orientation

It is not too difficult, as suggested in the Buchanan Report[1] to classify roads as essential or optional and there are no doubt several options in between. What is important is to clarify to motorists the nature of the road they are on and here the lighting designer can make a significant contribution.

There are several ways in which the road or pathway can be identified: by the illumination level of the light, the source colour, the height at which the light sources are placed, the road texture and its colour, the nature of lanterns used and the way in which they, and the light they provide, all relate to the surrounding buildings.

The nature of lighting streets and pathways which form a part of the lighting designers brief, must be seen as a part of the overall visual perception for the town – the visual masterplan.

Such a masterplan might well suggest that all main traffic routes be identified with a high level of cool light, while warm sources are reserved for minor routes and pedestrian areas. But, however this is planned, it should be coherent and lead to easy understanding by the public, whether on foot or by car.

What seems clear is that the street light which tries to cater for both cars and pedestrians using a single type of lantern, is more likely to fail. Examples now exist of tall columns along main traffic arteries designed for a high level of road lighting, having separate smaller fittings attached at a lower height to provide the more gentle and varied light for the slower moving pedestrian. Whilst such a solution may initially cost more, it is more functional, in that the main lighting can be directed towards solving the problem of the fast moving traffic, whilst at the same time the visual impression of the whole will be much improved for the pedestrian.

It should be possible for people to orientate themselves immediately and to judge the hierarchy of the road system they are negotiating, and this may in turn have an effect upon the speed of traffic.

Speed

It is universally recognized that speed kills and short of separating vehicular traffic from pedestrians, there is no perfect solution.

---

[1]Buchanan, C. (1963) *Traffic in Towns*, UK.

Methods of traffic calming can make some contribution, and perhaps the lighting designer may have a part to play in ways which have not yet been devised – 'black spots' emphasized by changes in the manner or colour of the lighting? The future will no doubt provide new answers, in which the lighting designer will need to play a part.

SPACES BETWEEN

These are the open areas between buildings not taken up by roads, which provide the welcome interludes . . . the parks and open spaces, described as the 'lungs of the city'.

The following may be regarded in this category:

The grand scale

- The city centre square or plaza. Examples: St Mark's Square Venice; Trafalgar Square, London.
- The large city park. Examples: Central Park, New York; St James's Park, London

The small scale

- The small town square. Examples: Commercial property around it; residential property around it; church yards
- Shopping precincts
- The village green used for games during the day
- Sports fields – local or national

These are the main types of open space although others such as the banks of rivers and canals should also be included. They can become an embarrassment, or be planned to provide a useful amenity for the town, and much will depend upon how they are perceived at night. During the day some may be full of people and interest, some used for open markets, some just for people to chat or play boule; but it is in the manner of the lighting at night by which their overall success will be judged.

Many cities have been planned, or have grown without any plan – such as Corbusier's 'pack horse town' – without thought being given to such facilities, and yet it is these open spaces which are of special importance in providing the quality of city life. London is particularly well-endowed in this respect, with its parks and town squares, open areas which are always under threat from those who wish to fill them with buildings.

DP Archive

Nighttime view of Nelson's column (Lighting designer Phoenix/Large)

The areas involved may be extensive, in the manner of central squares or town parks, or small pedestrian areas; but whether large or small, they are valuable assets, which need to be managed carefully so as to contribute to the overall appearance of the environment, and they must not be left to 'wither on the vine'.

Open spaces are where the pedestrian is king, and where the lighting must be at the human scale. The creation of safe pedestrian areas is the province of the town planner, to ensure the interrelationship of routes, footpaths and transport systems; but the lighting designer has a strategic role to play in ensuring that the spaces work in a functional sense at night; but far more than this, that the overall environment provides that added sense of harmony with the surroundings which can lead to delight. It is sad that so many open spaces are lit in such a way, as they do little to encourage their use by members of the public at night, who may even perceive them as threatening. This is not to say that open spaces have necessarily to be lit to high levels of intensity, or even at all, so long as the needs of safety and security are met – there should always be places for people to enjoy the dark and look at the stars.

Each open space will need analysis as to its purpose and function, in order to determine a lighting brief; each should be conceived as a national or local asset, according to its location to ensure that it contributes to an overall sense of relaxation, and ultimately to a sense of harmony.

INCIDENTS

Incidents can vary in size from the very large to the very small. They may be huge National monuments such as the Statue of Liberty symbolizing the gateway to the New World, the Eiffel Tower in Paris, Nelson's Column, and perhaps now including the 'Angel of the North' or the 'London Eye' in England.

Alternatively, they may be quite small: the 'Buttercross' in a market town, the fountains in Trafalgar Square, or a local war memorial or other item of civic sculpture. But whatever their size they should be regarded as lighting opportunities, as they can add significantly to the symbolic importance of a space at night and add flavour to an environment.

When discussing lighting incidents one should not forget those of an ephemeral nature, designed to make a statement perhaps for half an hour as a fireworks display or for a festival, for example at Christmas where they may be available for only a week or so.

There are also the *son et lumière* displays which, like Jean Paul Jarre's famous event *Lighting up Docklands*, can arrest the attention and inspire further exploration – the city makes a splendid backcloth against which to stage such extravaganza.

Courtesy of Philips Lighting

Nighttime view of Buttercross in a Market Town

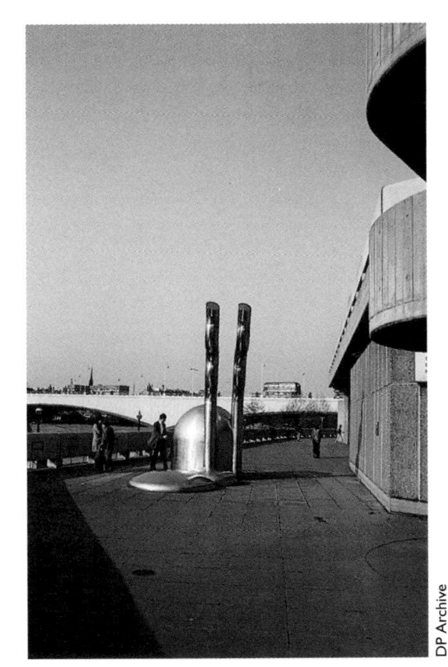

DP Archive

View of the William Pye sculpture by day (Designer William Pye)

DP Archive

View of the William Pye sculpture by night (Designer William Pye)

LIGHT POLLUTION

While not so much a feature of the visual master plan, as of the tactics which it may generate, the question of light pollution needs to be addressed as an aspect of the overall strategy applied to lighting the city – the question of glare has already been mentioned and the two are interrelated. Light pollution, or as it is sometimes referred to as light trespass, occurs because of the lack of control of light, allowing it to escape in unwanted directions, directions which may be the immediate cause of glare, resulting in the secondary cause of light pollution.

Street lighting is an obvious example, where the engineering of the lanterns is such as to allow the escape of too much sideways or upward light, often aided by a highly reflective road surface, so that high levels of direct and reflected light are transmitted upwards to the sky.

Buildings can also contribute towards light pollution, both from the escape of the interior lighting, and from exterior floodlights directed upwards, where insufficient control of the beams allows spill light to the sky. It is true to say that light pollution derives principally from poorly engineered lighting installations, and the associated glare they generate.

It is argued by astronomers, and others who watch the skies at night, that light pollution makes their task more difficult; this may be so, but an equally worrying aspect is the waste of energy with which it is associated. Anything that can be done by greater functional control of light should be done; this will have the added advantage of lowering the levels of light pollution at the same time as saving energy.

VISUAL MASTERPLAN

No two towns or cities are alike, and what is right for Athens would differ widely from what would be appropriate in Durham . . . but what is right for all, is that ideally there should be a visual masterplan developed to take in the transport systems, the roads, pedestrian ways, the open spaces, monuments or incidents and where appropriate the waterways and bridges – indeed, all the elements making up the lit environment of a place.

Buildings form the most important elements of a city, but the appearance of the buildings, their approaches and setting, is influenced by the roads and path patterns and their associated street lighting . . . the design of all of these elements are at the heart of the visual masterplan. The drawing-up of such a plan should be done by a team comprising architects, planners, landscape architects and others including highway engineers and lighting designers; although not simple, it is not impossible. The most difficult problem of all is not so much the drawing up of the plan; it is the determination and political will to carry it through.

Courtesy of Ville de Lyon

A general view of an area of the Rhône and the town of Lyon

In London where, until 2000, there was no overall City Council, no overall transport policy, and where every decision had to be taken at the local level by 32 different boroughs it might have seemed impossible – even getting a masterplan drawn out would be one of the labours of Hercules. In a new town the problems are at their easiest, provided the need is recognized at the inception.

The more general problem is in our older towns, like Durham, where much can be done, and a masterplan created. There will never be a final solution because a town must be a dynamic structure, changing to meet modern needs; but it is these very changes which can assist, for wherever a change occurs, or areas need to be renewed, they can be related to the broad objectives of the overall masterplan, allowing the visual pattern to

Courtesy of Ville de Lyon

A courtyard in Lyon with a lit fountain

develop . . . a plan which will develop at an increasing rate once the advantages of its guidelines can be seen. The aim must be to construct towns where there is a vibrant living nightscape, and where every prospect pleases.

A good example of the advantages of a master plan is that of the French city of Lyon. Over the years a partnership of professionals, together with those in business who can influence such matters was formed to revitalize the nightscape of the city, with spectacular results. A walk around Lyon at night is an experience from which much can be learned. The Place Terreaux in Lyon (see Chapter 4, Incidents, p. 74) features a sculpture of Bartoldi and horses which, at night, is illuminated by concealed lighting.

Trafalgar Square

The proposed lighting scheme for London's Trafalgar Square, drawn up in 1995, is a further illustration of the advantages of a visual masterplan; with elements of the proposals being incorporated in later schemes. Commissioned by the City of Westminster, proposals were put forward by architects and lighting designers for the lighting of the square, the fountains and Nelson's Column, together with the surrounding buildings. The designs were such that when detailed plans were submitted these would conform to the overall concept.

To illustrate their approach the consultants provided coloured visualizations of the various aspects of their proposals, some of which are illustrated here.

Three perspectives of Trafalgar Square, London (b and c on following page)

(a) Nelson's Column (Lighting designer LDP)

(b) Elevation to the National Gallery (Lighting designer LDP)

(c) The fountains in the square with St Martin's in the background (Lighting designer LDP)

All these aspects contribute to the concept of a whole lit environment for the area, conforming to a Visual Masterplan. This is an example of where the plan is confined to a limited area, but the concept can be applied to an entire historic centre of a town, as it was in Edinburgh.

Trafalgar Square as it is in 2001 is illustrated as a lit environment (p. 130) and it is interesting to see the correlation between the visualizations depicted in the Masterplan illustrated in this chapter and the result which can be seen today; it is fair to say that had the masterplan not been drawn up, the resulting piecemeal development would have been far less satisfactory.

2 Buildings

Analysis ... initial decisions ... solid form, external floodlighting ... glazed form, internal lighting ... mixed form, internal/external lighting ... roof form, translucent/reflective

ANALYSIS

As the most important element in the night appearance of our towns and cities, the way in which buildings contribute to nightscape must take precedence over the spaces between and the incidents which make up the rest.

While it might appear logical to analyse buildings in terms of their function – churches, offices, shopping, housing and so on – there is too much similarity between the methods of exterior lighting used for widely different architectural programmes, whereas the form of each building is the main factor which differentiates the approach to be used.

The methods of lighting themselves will be discussed in more detail in Chapter 5, Tactics. Our concern here is with the exterior appearance of the building at night, rather than detailed discussion of the light sources involved.

The following 'building forms' encompass the majority of buildings:

1. Solid form	External floodlighting
2. Glazed form	Interior lighting
3. Mixed form	External/Internal lighting
4. Roof form translucent/reflective	Internal lighting

Before discussing the approach required in terms of the above forms of building, it is useful to identify the decisions which need to be made, in order to establish the brief for the building which must be discussed and agreed with the client.

Decision 1. Should the building be lit at night?

Where a visual masterplan of the area is available, as is the case in an increasing number of our towns, reference should be made to this to establish the relative importance of the building or groups of building in their location and their role in the nightscape.

This is the most important decision of all, for there is no imperative to light up a building at night. Indeed, many buildings at present floodlit would be better left in decent obscurity, or lit from their own internal lighting. It will be up to the client together with his advisors to establish a need, and a justification; this is particularly the case with sports facilities, for the addition of what might be judged to be an intrusion in a neighbourhood.

The type and nature of the building in its location will establish whether exterior floodlighting is desirable, or whether a different aesthetic is called for. Having established the need for an exterior lighting scheme, and since this will have repercussions, it will be necessary to hold discussions with the planning authority who will wish to consider the visual impact of the proposals on the area. At a later date they will be involved in agreeing the type of equipment, its location and questions of light pollution, and likely objections.

Decision 2. Relationship with daylight appearance

It is important that there is a correlation between the view of a building seen by natural light during the day and by artificial light at night and this needs to be considered in two ways: first the desirable appearance of the building itself at night, and second the appearance of the lighting equipment, the luminaries, since obtrusive equipment in ill-considered locations may or may not provide a satisfactory light at night, but can easily ruin the appearance of a building during the day.

A visual study of the building during the day, seen in its context within the town, needs to be made to establish whether it is desirable to attempt to light the whole building, or whether certain elements would be better left unlit. The lighting architect is in the difficult situation of making decisions about the appropriate appearance of a building when in many cases the original architect will not be around to give his views. A case may be made for giving the building a variety of appearances, a change with the seasons, or times of night.

For the architect the most important aspect must be the question of architectural 'unity'. The concept of unity would include the important aspects of modelling, colour and scale, all aspects on which there will be divided opinions; opinions which may reflect the needs of fashion, and change with time. The unity of a building at night depends almost entirely on the manner of its lighting both interior and exterior; it will be the architect's responsibility to ensure its achievement.

The appearance of the building during daylight hours should be studied as a reference point; for although the building should be seen to be the same building at night, its appearance may be modified to emphasize its good points, while excluding its less desirable aspects.

We may be on dangerous ground here in those cases when the original architect is not around, and the decision as to whether the lighting designer should play God is critical in establishing the brief with the client. One thing is certain, the influence of an architect's view is essential; perhaps the reason why some of the most successful projects have been directed by lighting designers with an architectural background.

Decision 3. The desired appearance of the building in the nightscape

Careful consideration must be given to establishing the way in which a building should be experienced at night and this is the essential role of the designer. The façades of older buildings are often rich in decoration, with windows, cornices and columns providing modelling – where the three-dimensional quality of the façade can be emphasized by careful floodlighting. The more simple façades of modern buildings are less amenable to exterior floodlighting at night; but are often such as to achieve both interest and unity, without the need for extraneous decoration and have less need for floodlighting.

Decision 4. The effect of the location of the building

The context in which the building will be seen needs to be understood. There is a very great difference between how an isolated church in a country setting is experienced, where there is little or no other environmental light, and the building in a city centre.

This relates not only to the light level suited to its location – the difference between the light required for a village church in a low key situation, as opposed to that of a city centre – but also to the direction from which views of the building can be enjoyed.

There will be buildings where views from all sides cannot be obtained. There will in some cases be viewpoints which close a vista, define an important side to a square, or in some other way are more important than others. For this reason, if not for reasons of economy or limitations of lighting technology, it will generally be necessary to establish a priority of importance of the views of a building to be emphasized at night.

It must also be remembered that the nature of the materials of the building, whether light or dark, must be taken into account, as more light will be required on dark surfaces to achieve the same effect as that on a light surface. This leads on to the question of colour.

Decision 5. The influence of colour

An aesthetic judgement needs to be taken as to the relationship between the colour of the building, and the colour of the proposed light source. As a general rule it is better to emphasize the natural colour of the building, by the use of a comparable light source – a white light source for a white building material, such as concrete or stucco, and a warmer source for red brick or warm coloured stone. Having said this, the introduction of light sources with improved colour characteristics makes such distinctions less necessary. It is no coincidence that during the recent Millennium Church Flood Lighting Programme, most churches were lit with the latest Metal Halide (CDM) lamps, available in cool or warm colours, designed to bring out the natural colours of stone or brick.

With the introduction of the HP Sodium lamp in the 1960s it appeared to be easy to flood whole areas with this warm light source, in the mistaken view that all buildings look better at night when warmed up – they do not.

A recent tendency has been to use coloured light, and even colour changing systems, on buildings, and a case may well be made for this in special circumstances of ephemera; but careful thought needs to be given to this to avoid bizarre effects.

The colour appearance of a building lit at night has two main components: the colour of the bricks, stone, concrete, or metal of which the building is formed, and the colour of the light source. While the colour of the former is unlikely to change throughout the life of a building, the influence of the light source in determining its colour appearance may be changed radically.

It may change because new sources of cool or warm colour are available; or because the means are available for the colour of the light source to be changed at will, or sequentially. The art of *Son et Lumière* developed in the 1960s, where the history of an old building might be told by a theatrical display of lighting added to the designer's repertoire, in showing the effects which can be created by coloured light; lighting which changes to suit a mood or illustrate a story. Almost anything is possible, and this is really the danger – in the wrong hands, the emotive variety of the colour now available can be disturbing – or as Mies Van der Rohe said in another context 'I find your variety very monotonous'.

Decision 6. Location and appearance of equipment

Despite the fact that the overall development in light sources is for miniaturization, the exterior equipment needed to floodlight a building tends to be bulky, and an initial assessment should be made of the most suitable locations for mounting the floods, where they will not lead to glare and light pollution at night, or visual obtrusion during the day, while being accessible for maintenance without undue difficulty – often a conflict with the possibility of vandalism.

Decision 7. Lighting trials

While there are now methods using computer generated graphics, to give a client an indication of the final lighting effect on their building and, while artists can also produce perspectives which give a reasonable impression of a lighting scheme, there's nothing quite like a site trial of the recommended lighting equipment itself, to show a client what it will be like 'for real'. This gives the added advantage that alternatives of colour and direction of light may be tried.

Each of the four forms of building identified earlier, need to be discussed with examples to show the approach to the nighttime lighting which will suggest itself.

BUILDING FORM

Solid form

Until the structural revolution of the modern movement in architecture from the early 1900s, buildings were mostly solid, with apertures punched in the façades to admit daylight (originally known as wind-eyes from which the word window was derived). They were experienced for the most part at night by light escaping from uncurtained windows, or by means of light sources placed externally and directed at the building, which we know as floodlighting.

Examples of solid form

It is this type of building – where the solids exceed the voids – described here as 'solid form', which comprised the vast majority of the world's contruction until the early twentieth century; and which even today forms a significant part of building work.

The illustrations and diagrams shown here are of a series of buildings of this type. The examples have been chosen to show the wide variety of solid form in which the type and colour of light is manipulated to provide an appropriate night appearance – from the highly decorative façade to the more simple.

Leeds Castle, Kent

Surrounded by an artificial lake, the fourteenth century Leeds Castle in the Kent countryside has the advantages of a low key background allowing comparatively low levels of lighting to be very dramatic, and the benefit of water with reflections of the castle in the lake adding a further dimension. The lighting scheme is an economic solution, in which five 1000 kW metal halide lamps are located in embrasures set into the banks above water level. Since the castle can be viewed from all sides a degree of compromise had to the made, since it would have been impossible for flood lights on one side not to be seen from the opposite bank. However the way in which this has been done is designed to minimize glare, and to allow unimpeded views from the main approaches to the building.

It is not only historic buildings which may have solid form. Leeds Castle has the unsymmetrical pattern of windows punched in the solid stonework characteristic of mediaeval fortified architecture; while the form of Frank Lloyd Wright's Guggenheim Museum in New York

DP Archive

Leeds Castle, Kent, night view (Lighting designer Thorn Lighting)

illustrated earlier (p. 7) typifies a modern example. From the point of view of their nighttime lighting, the methods adopted may differ widely, but the quality of the night appearance is not dissimilar.

Worcester Cathedral, Worcester

This eleventh century building in the heart of the ancient cathedral city of Worcester, is perhaps the prototype building illustrating the need for a

Courtesy of Thorlux Lighting

Courtesy of Thorlux Lighting

Worcester Cathedral. Exterior pictures at night

View 1: Distance view
View 2: Distance view across water

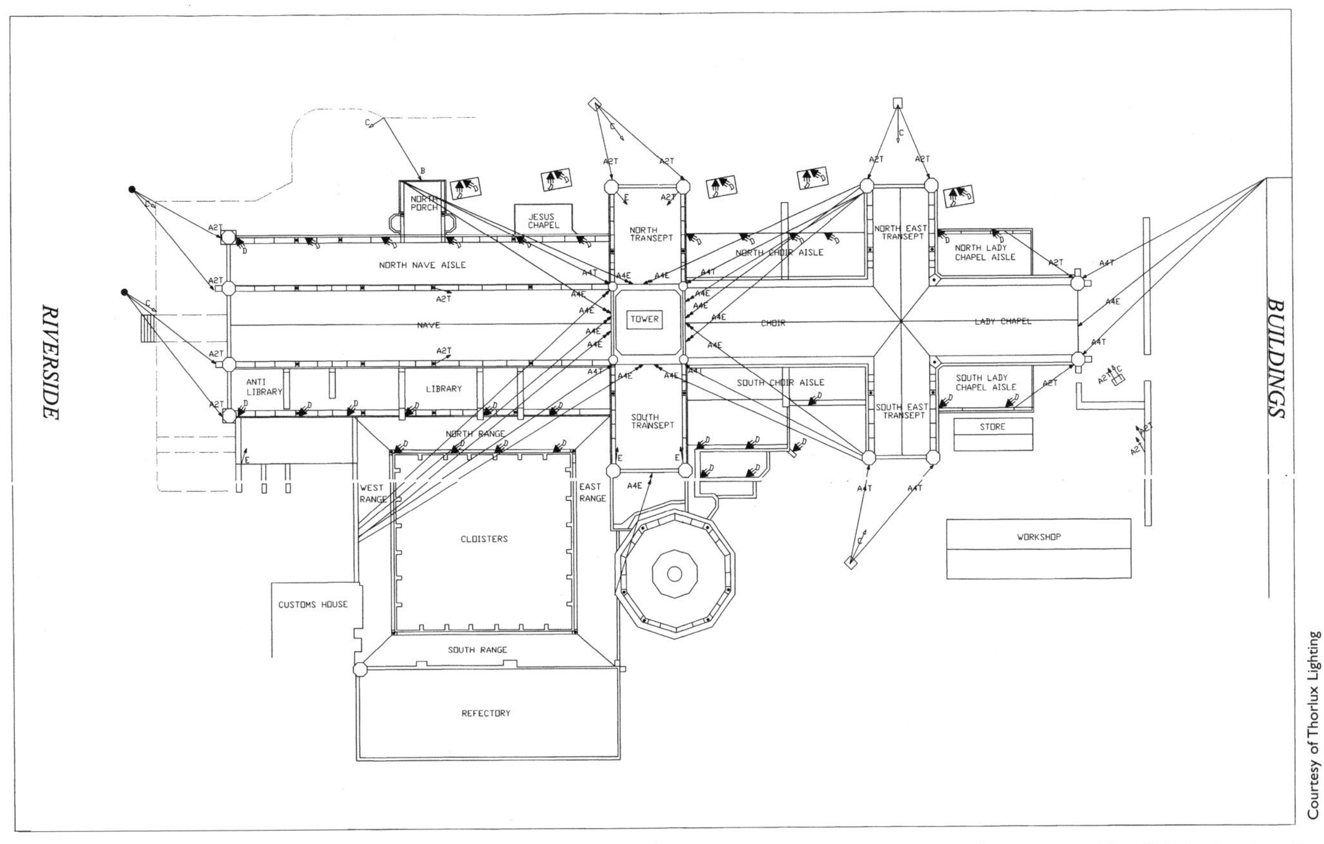

Courtesy of Thorlux Lighting

Plans to show proposed floodlighting location of Worcester Cathedral

scheme of external floodlighting, if the building is to register by night. Should a visual masterplan have been available for the city, the cathedral would have been at its heart. In the event no such masterplan existed, and it is a tribute to the designers that the final result would fit well into some future lighting framework for the city.

The scheme of floodlighting, with its warm colours echoing the warmth of the stonework, brings out the solid form of the cathedral, both at close quarters, and when seen at a distance. The lighting equipment has been located and concealed with care to avoid glare to the public, and light trespass to neighbouring buildings, while providing a three-dimensional quality to the different faces of the building, with proper emphasis being given to the more important façades.

The Unilever Building, London

This prewar commercial building on the side of the Thames at Blackfriars has a simple classical façade, with heavy modelling of cornices and columns. Floodlights at ground level are concealed from public view by a surrounding wall, with floods concealed at balcony and roof level to light both the upper storey, and a line of sculptures well-modelled by low voltage directional floods.

The proposals were tested on site by trials of a variety of colour filters (20 different colours were tried before the final selection was made). The whole appearance is controlled by a Lutron control system which changes

The Unilever Building, London. Lighting trials to columns (Lighting designer Graham Large)

The Unilever Building, London. Daylight view of the façade (Lighting designer Graham Large)

The Unilever Building, London. Nighttime view of the façade (Lighting designer Graham Large)

both the colour and intensity so that passers-by are continually surprised by a slightly different appearance to the building on a daily basis.

Cour Carrée, the Louvre

The night lighting design for the square court at the Louvre, is an example of the use of computer simulation to create an impression of the nighttime appearance of a building to obtain the agreement of the client. Both the method of design, and its implementation is a very hi-tech approach, and would not be applicable to the average floodlighting scheme; the cost involved would prohibit this as a general approach to nighttime lighting.

Having established the visual impression of the modelling of the façade required, special equipment was manufactured to be placed below the cornices to achieve this effect, the appearance of the modelling of the

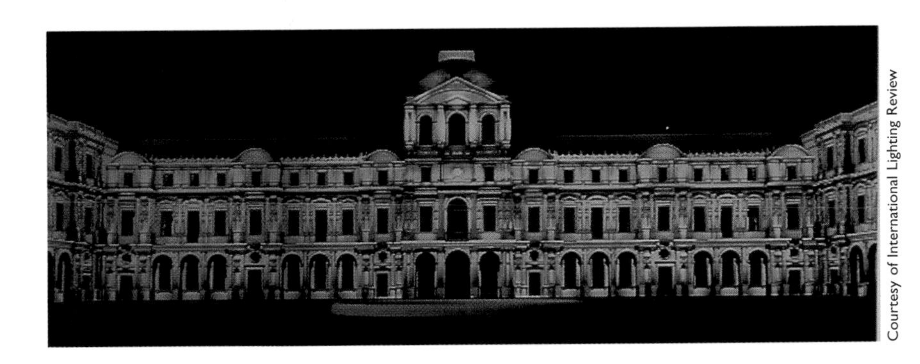

Courtesy of International Lighting Review

Cour Carrée, the Louvre. Computer simulation of night appearance (Lighting designer Philips Lighting)

Courtesy of International Lighting Review

Cour Carrée, the Louvre. Actual night appearance, on completion (Lighting designer Philips Lighting)

façade bearing this out. The angle of the light is closer to that of sunlight, than is usually found from the upward distribution of light from traditional floodlighting.

Palace of Fine Arts, San Francisco

Built in 1915 for the Panama Pacific Exhibition, the monumental palace designed by the architect Bernard Maybeck had never intended to be permanent, and by 1962 had fallen into disrepair. It was reconstructed and reopened a few years later, but once again despite some efforts to revive it by a scheme of lighting, it fell prey to vandalism and lay unlit and derelict for a further 20 years.

It was not until a team of dedicated citizens formed a 'Light up the Palace Committee' that money was raised and the lighting designer Ross de Alessi was able to put forward the imaginative lighting design for the Palace which we see today.

In the eighteenth century it might have been called a folly, in that it has no specific function, but to be itself; the lighting is very low key, suitable to an area which is mainly surrounded by housing. The improved colour high pressure sodium light sources provide a sufficient amount of warm light from below using sources concealed in precast concrete containers, with other sources at roof level, providing what the architect had intended as a graze of light to enhance its sculptural appearance.

Courtesy of Doug Salin

Palace of Fine Arts, San Francisco, close-up view (Lighting designer Ross de Alessi)

Courtesy of Doug Salin

Palace of Fine Arts, San Francisco, long distance view (Lighting designer Ross de Alessi)

Copenhagen Airport, the Rotunda

This might be described as a piece of sculpture to conceal a functional electrical switching station, but it comes into the category of solid form building, despite the fact that it is formed of glass panels – panels which require no exterior lighting, since they are lit by integrated internally illuminated remote source lighting fibres, making the Rotunda glow at night.

The Rotunda Copenhagen Airport at night (Architect A Gottlieb & Paludan; Manufacturer Louis Poulson)

The architects, Gottlieb & Paludan cooperated with the artist and sculptor Ingvar Cronhammar to design the Rotunda, an artist well-known for large sculptures having dynamic impact. The circular Rotunda is therefore both a solid and a glazed form, the vertical slot being the main characteristic of the circular façade, set against a glowing exterior in which the remote light sources are placed at high and low level, where they can be maintained easily.

Cranhill Water Tower, Glasgow

The tower is one of 10 water towers built around the outskirts of Glasgow, and might be thought of as a rather unusual form; but it falls easily into this category. The project, standing 25 metres high was a community project for the town during Glasgow's reign as the UK City of Architecture and Design, in order to create a public space at night with a permanent lighting installation to serve the immediate community.

The lighting concept developed by the architect Adrian Stewart, was to wash blue light up to the central soffit, while green light would reflect off the perimeter onto the supporting columns. Orange and red filters are used inside the stair core, with light spilling on to the soffit. The tank area is largely unlit, except for the projecting fins which have narrow blue beams to emphasize their presence. A lantern at the top glows orange in contrast. Lighting has transformed what had been perceived as a rather sinister shape lurking in the dark, to an exciting lit incident for the inhabitants of the area.

Plaza de Toros, Valencia

A seminal example of the exterior lighting of a solid form building, by using the window embrasures. The detail of the light fitting used

Courtesy of Adrian Stewart

Cranhill Water Tower, Glasgow, night view distance (Architect CH Stewart)

Courtesy of Adrian Stewart

Cranhill Water Tower, Glasgow, night view close (Architect CH Stewart)

indicates a simplicity, which in no way detracts from the appearance of the building, putting the light where it is required around the window embrasure, avoiding light pollution, and light trespass. No other exterior lighting is provided for the main structure, but to give clear definition to the top of the building there is added lighting at roof level, which completes the overall lighting appearance.

The method of lighting used is one which would be used to show off the building as an historical monument, being unrelated to its use, and therefore savings in energy could be made by lighting up the building only on certain occasions, i.e. for reasons of tourism.

Plaza de Toros, Valencia at night (Manufacturer l'Guzzini)

The Historical Museum of Catalonya, Barcelona

Glazed form

This is perhaps the simplest form of all, made possible by the structural revolution of the modern movement. Here the window forms the entire façade, whether at single or multi-storey level. The appearance during the day, where the façade tends to appear dark, is converted at night into a 'peep show', with the interior glowing with light.

If the building is a row of shops then its purpose is to expose and display the wares. If an office, where privacy may be of some importance, there may appear to be a need for some means to be adopted, such as blinds, to obscure the interior. What is clearly of importance is the appearance of the interior lighting equipment seen from the outside.

The interior lighting system for any building will form a part of its nightscape, but where the building is of glazed form, as for example in the Manufacturer's Trust Building (see p. xvi) it will play a major part, and consideration must be given to its appearance, often seen as conflicting at higher levels in multi-storey buildings.

In buildings of glazed form exterior floodlighting is a positive disadvantage. It is often forgotten that the advantage of the easy view-in to a glazed building comes with the responsibility to ensure that the view is worth looking at, and not a collection of the backs of filing cabinets or piled up furniture. A good example of a glazed building where the interior view is immaculate is that of the Orange Operational Facility, by architect Nick Grimshaw; this is in stark contrast to those where this factor has been forgotten.

Where the building has an enclosed space within, such as a theatre, surrounded by an outside perimeter of glass, it is the interior wall which will be experienced from outside, and it is this that must be flooded with light if the interior space is to register. This can only be done internally, and again exterior floodlighting is unwelcome.

The illustrations and diagrams of a series of buildings which follow are in this category, and again the illustrations have been chosen to capture the wide variety of form, showing the importance of the interior and the manner of the interior lighting.

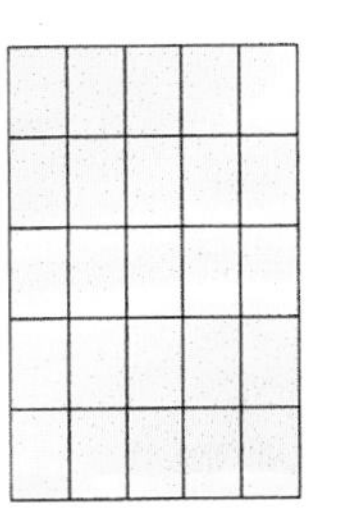

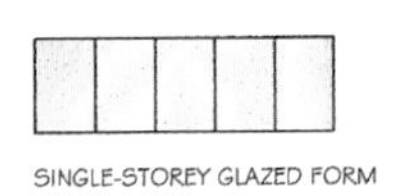

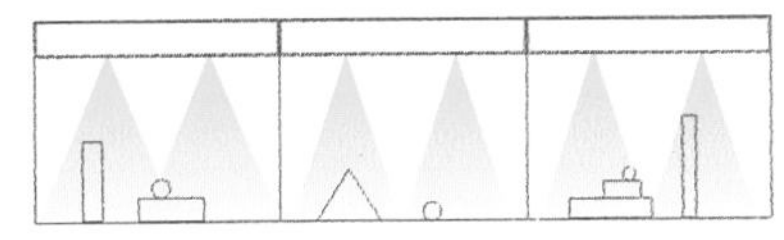

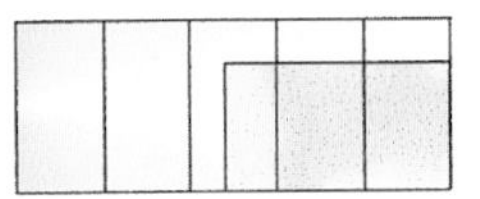

Examples of glazed form

Courtesy of Les Shipsides

Orange Operational Facility, night view (Architect Nicholas Grimshaw & Partners)

Willis Faber and Dumas, Ipswich

The offices of Willis Faber and Dumas in Ipswich built by Norman Foster in 1975 were perhaps one of the first buildings in the UK to illustrate the extreme nature of the glazed form.

During the day the perimeter of the WFD building – reaching to the edge of the site and following its contours – appears black, heightened by the use of a dark modifying glass, while at night the interior lighting

Courtesy of Foster & Partners

Willis Faber and Dumas building, Ipswich (Architect Foster & Partners)

shines through to define the interior spaces. The appearance of the ceilings of the building, which is three-storeys high, becomes visually more important at the higher storeys.

The interior lighting has been designed to provide a unifying factor for the higher ceilings around the perimeter; this has been made possible since the building has been constructed for a single client. In taller office buildings where different floors are often leased to different companies, there is a danger that each company may wish to express its own individuality, leading to a degree of anarchy at the different floor levels.

In the early days of fluorescent lighting, the nature of the building, whether as a single entity or speculative office, could be discerned by the conflict in the nature, direction and often the conflicting colours of the lamps in the light fittings. This criticism can of course be levelled at any type of accommodation, but is particularly apparent with the glazed form making it all the more necessary to consider the second aspect of architecture in this respect.

The all-glass façade has now become common, from the glass shopfront to the airport terminal – the terminal at Stansted being an example by the same architect, a few years later.

Roseberry Avenue

The office building at Roseberry Avenue by architect John McAslan is typical of the majority of glazed form buildings; where, despite the floor spandrels being identified, the area of glazed frontage dominates the façade, rendering any attempt to floodlight the building from the front inadvisable.

In this four-storey building, built for a single owner, the interior lighting of recessed linear fluorescent fittings running parallel with the window wall provides an ordered solution. The type of light fitting, using low brightness louvres giving no spill light to the ceilings, reduces any glare to the immediate area beyond the building. The glazed staircase at the corner adds a desirable verticality to the façade.

One might expect that with a fully glazed front, there would be a degree of spill light to the immediate exterior perimeter, but here, because of the nature of the interior lighting design, it would be necessary to light any footpath around the building independently.

Helsinki Opera House

The Opera House in Helsinki, by Finnish Architects Eero Hyvamaki, Jukka Karhunen and Risto Parkinen illustrates the type of glazed form where the interior, the concert hall itself, is a solid form within a glazed exterior.

Here it is the lighting of the solid form which can be seen from outside; while the exterior glazing is transparent permitting views at night to the interior from outside and alternatively during the day, views to the exterior countryside from the foyer inside. Therefore, this is the form that requires no light to be thrown on it from the outside, the solid form inside being lit by internal light sources directed towards it. The opera house clearly demonstrates the way in which the internal structure is experienced through the transparent glass enclosure.

Courtesy of Peter Cook/View

Roseberry Avenue, night view (Architect John McAslan)

Courtesy of Erco Lighting

Opera House, Helsinki, the interior (Architect Hyvamaki, Jukka & Risto Parkinen; Lighting designer Joel Majurinin Ky)

Courtesy of Erco Lighting

Opera House, Helsinki, the exterior (Architect Hyvamaki, Jukka & Risto Parkinen; Lighting designer Joel Majurinin Ky)

Imax Theatre, Waterloo, London

Similar in form to the Helsinki Opera House, though widely different in context, the Imax Theatre in London is designed to show both 2-D and 3-D films. From outside, the auditorium experienced through the exterior glazing makes it unnecessary for any additional exterior lighting.

The auditorium has a large wrap-around Howard Hodgkin mural to its curved surface, which is experienced at night with vivid colour, making it a significant landmark in what was once a rather run down area of the city. While the exterior glazing to the theatre is a permanent aspect of the building, it would be expected that the interior wall design might change many times during the life of the building.

DP Archive

The Imax Theatre, Waterloo, night view (Architect David Hersey Associates)

Law Courts, Bordeaux

The French Ministry of Justice requested the architect, the Richard Rogers Partnership, to provide a building that would emphasize by its transparency a more positive perception of the accessibility of the French judicial system; so it is appropriate that here, like the Opera House in Helsinki or the Imax Theatre, the exterior would be a glass skin, allowing views through to the solid Courts beyond.

While this is less apparent during the day when the glass surround appears dark, at night the transparency of the exterior skin allows views to the Courts, lit from the interior.

The wide variety of 'glazed form' buildings, whether seen directly or 'through a glass darkly' form an increasingly significant part of the urban nightscape.

Courtesy of Christian Richters

Law Courts, Bordeaux, night view (Architect Richard Rogers Partnership; Lighting designer Lighting Design Partnership)

Mixed form

These are buildings where areas of solid wall are counterbalanced by glazed areas. Since the 1960s, a large majority of city buildings are of the mixed form type, or buildings where neither the glazed form nor the solid form dominate. The form of the building will have been developed from the functional needs of the plan, but where the second aspect will have informed the architect's design, and where the exterior appearance at night will result from a combination of views into the space with its interior lighting scheme, associated with areas of blank wall which may or may not be lit by floodlighting from the outside.

Mixed form is illustrated in terms of lighting by the diagram, from which it can be seen that the combined types of form, glazed or solid, are identified. The appropriate type of lighting for each form has already

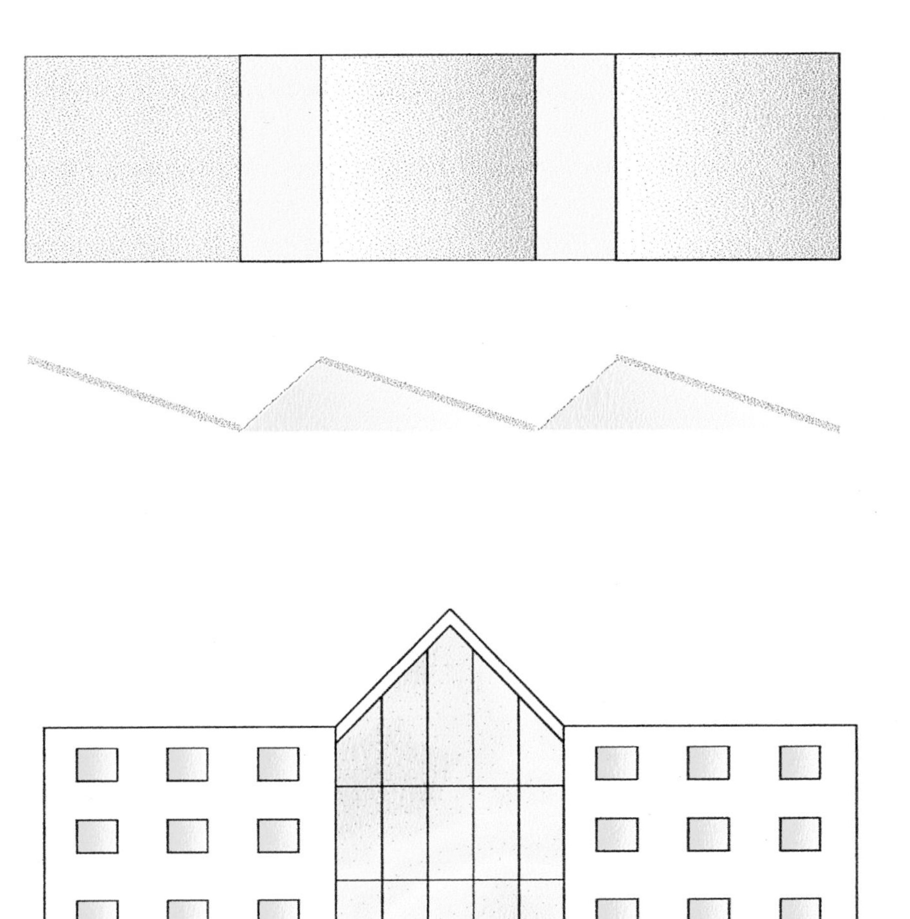

Examples of mixed form

been discussed, which broadly means that the solid form may be left unlit or lit by exterior sources, and the glazed form by its own interior lighting. However, discussion of the following projects shows the enormous variety of approach available to the architect.

The Guggenheim, Bilbao

An example where the form of building allows the solid elements of the exterior to be lit by carefully related glazing from the interior, is the new Guggenheim art gallery in Bilbao by Frank Gehry. This might at first sight appear to be a solid form building, but a more careful assessment reveals its sophistication.

More a work of civic sculpture than an art gallery, the counterpoise of light and shadow on the metal titanium surfaces at night expresses delight, and epitomizes the mixed form, where the light from inside plays on the exterior surfaces. This is not to forget the sculptural effect of daylight and sunlight on the curvilinear forms, an experience which echoes that at night, ensuring that the day and night appearance has an equivalence – clearly the same building. It is its aspect as a civic sculpture which has had a huge impact on the life of Bilbao, converting it from a

The Guggenheim Museum, Bilbao, exterior view of the 'Fish gallery' (Architect Frank Gehry)

The Guggenheim Museum, Bilbao, view from the bridge (Architect Frank Gehry)

rather indifferent ferry port to a must-visit town, where the exterior nightscape has significantly contributed to its success.

A classic example of the mixed form is a building set on the side of the lake at Stockley Park, a high-tech Business Park close to Heathrow Airport. Various elements such as the staff restaurant, and the entrance hall are glazed, while the main elevation formed of solid white panels is lit to a low level from exterior sources.

Night of the Hasbro Building, Stockley Park, from across the lake (Architect Arup Associates; Lighting design DPA Lighting Consultants)

Night of the Hasbro Building entrance, Stockley Park (Architect Arup Associates; Lighting design DPA Lighting Consultants)

Royal Library, Copenhagen

This is an extreme example of the mixed form, in that while the largest part is solid, with small slit windows appropriate to the work of the library, the large central atrium appears dark during the day, but glows with the interior lighting at night.

It is perhaps arguable whether the large solid areas of the façade would benefit from some exterior floodlighting from close offset floods, but the architect clearly felt that the exterior façade required no floodlighting, being identified by the horizontal slit windows. The interior glows out from the atrium, making a significant change in the overall appearance of the building after dark, but a change which is wholly appropriate to the building's integrity.

Courtesy of Bent Ryberg/Planet

Royal Library, Copenhagen, night view (Architect Henning Larsen; Lighting designer Louis Poulson)

Fitzwilliam College Chapel, Cambridge

The curved dark brick walls of this chapel contrast well with the large window behind the altar, which allows views to a large plane tree in the garden beyond. Seen from outside at night it is the large window with its interior lighting which dominates the view; with the curved exterior walls left in darkness, an entirely appropriate use of the mixed form. The views of the beautifully designed interior of the chapel are to be enjoyed from the chapel garden.

European Court of Human Rights, Strasbourg

The buildings, designed by the architects Richard Rogers Partnership was opened by President Mitterrand in 1995 and consist of the circular courtroom and commission rooms linked by a circular glazed entrance hall, together with public and press amenities; in addition, two cascading blocks contain offices for the commission.

Courtesy of Michael Evans

Fitzwilliam College Chapel, Cambridge, by day (Architect McCormac Jamieson Prichard)

Courtesy of J & J Services

Fitzwilliam College Chapel, Cambridge, by night (Architect McCormac Jamieson Prichard)

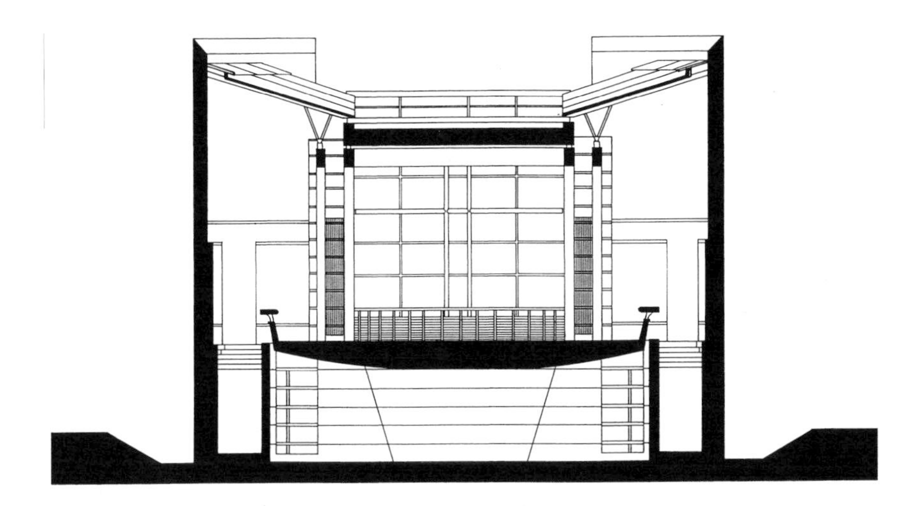

Plan of Chapel, Fitzwilliam College, Cambridge (Architect McCormac Jamieson Prichard)

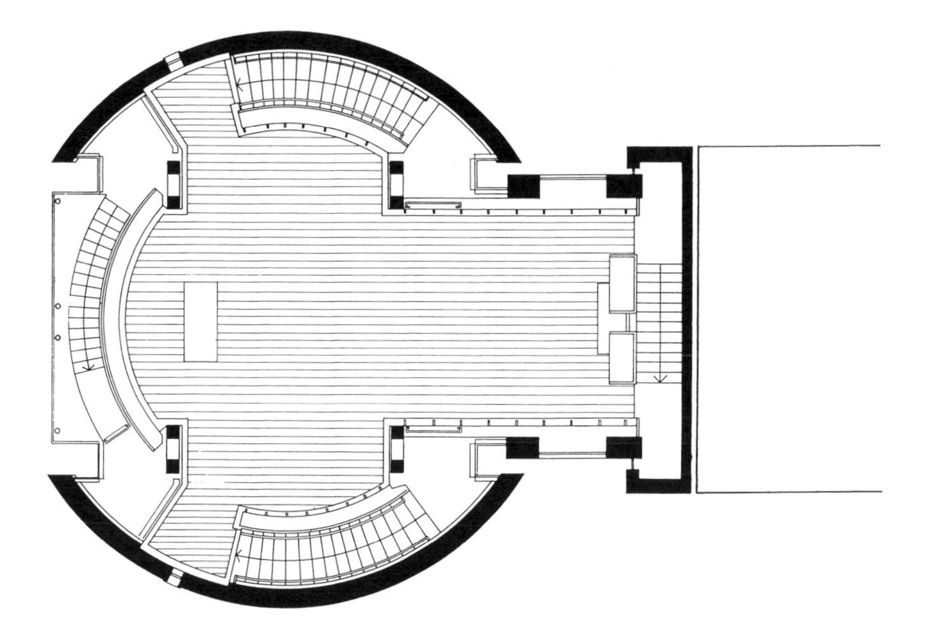

Diagram section of Chapel, Fitzwilliam College, Cambridge (Architect McCormac Jamieson Prichard)

It would be natural in such a large complex of mixed accommodation for there to be large glazed areas admitting daylight, and other closed areas with solidity – a mixed form.

UFA Multiplex-Kino, Dresden

This multiplex theatre in Dresden follows others in Hanover and Cologne described as reminiscent of a moon-rocket launching pad, exhibiting the same tendency as the glazed form of the Imax Cinema in London; in that it is the interior auditorium space that can be discerned from the outside, but in this case there are solid areas which create the contrast with the glazed spaces, giving the mixed form approach. The interior lighting is dominant as seen from the outside, with some additional and rather irrelevant light to some of the solid walls.

It is probable that the mixed form of building will dominate for new construction in the future, as it is already beginning to do so. For this reason the art of floodlighting will continue to predominate for historic buildings, but will tend to wither on the vine as far as it applies to the new.

Courtesy of Morley Von Sternberg

European Court of Human Rights, Strasbourg (Architect Richard Rogers Partnership; Lighting designer Lighting Design Partnership)

Courtesy of Christian Richters

UFA Multiplex-Kino, Dresden, night exterior (Architect Coop Himmelblau; Lighting designer Harry Hollands)

Roof forms: translucent/reflective

It is possible that the first building form of this type would have been the Buckminster Fuller geodesic domes of the 1950s, where the structure was formed of a framework permitting lightweight translucent coverings; alternatively, it could be argued that it was preceded by the early greenhouses.

The theory of translucent structures themselves may not have greatly advanced since this time, but the magnitude of the resulting structures certainly has, with the epitome being the Millennium Dome in London designed by Richard Rogers Partnership having a diameter of 400 metres.

Examples of translucent form

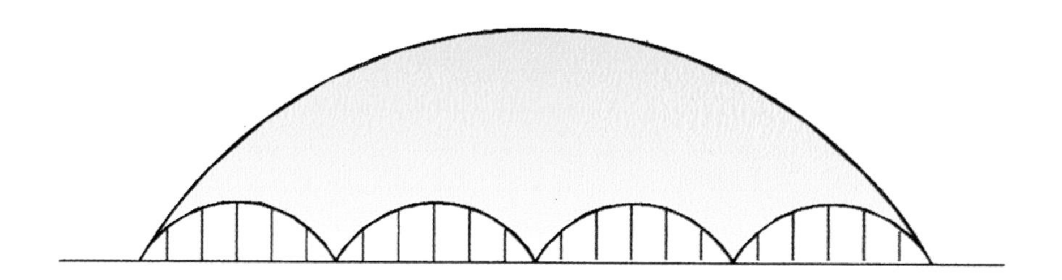

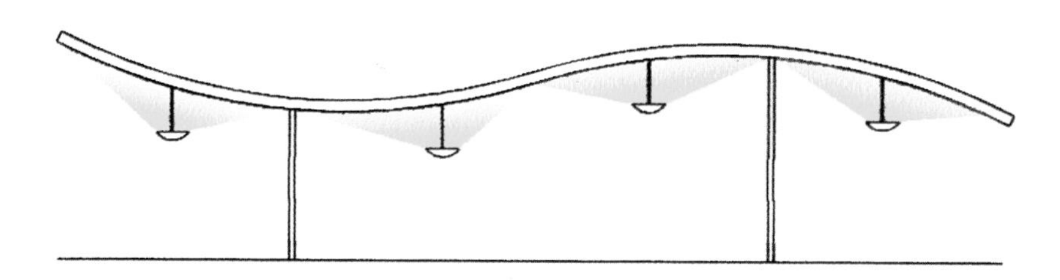

The Millennium Dome, Greenwich, London

The Greenwich Dome is the purest of translucent form in that it covers the entire exhibition site reaching to the ground all around, with views into the dome at ground level at selected points protected by glazing.

The translucent material of the dome acts as a giant screen, upon which can be played all manner of lit effects. The interior exhibits make little impression on their own from the outside, while the colours and patterns seen from outside can be altered at will by sophisticated forms of electronic control of projectors trained up to the inside of the roof.

For a building of this type there is no need for floodlighting from the outside; no need, but a possibility for display purposes.

From an exterior lighting viewpoint the fact that a translucent roof allows the interior light to escape to the outside may be due to a conscious decision to uplight the material giving a designed appearance

Courtesy of Grant Smith

The Millennium Dome, exterior view (Architect Richard Rogers Partnership; Lighting designer Speirs & Major)

to the roof seen from the outside, or alternatively to allow the interior lighting of whatever nature it might be to determine its night appearance. The former might be seen as a form of light pollution, whilst the intensity of the lit surface from the latter will be less.

Depending upon the nature of the structure supporting the roof and its height, the roof will either be the most important element of the lit environment, or will be associated with the surrounding support structure. It is likely that although this is a minor form when compared to the others already mentioned, it is one that may well expand in the future, and should be considered from the point of view of its impact on the lit environment.

The Pyramid at the Louvre

The structure itself may be of as much importance as the covering material as demonstrated by I.M. Pei's glass pyramid at the Louvre. Since the material in this case is transparent, the views out are as important as the views from within.

The Eden Project, Cornwall

A further example where the structure itself is of the greatest importance is the recently completed Eden project by architect Nick Grimshaw. Basically a giant greenhouse, it measures some 200 metres long by 47 metres high and consists of a series of biomes permitting daylight to the space. Normally white (see p. 42), a special installation emphasizes the possibilities of colour, for which reason it might be considered an ephemera (see p. 83).

DP Archive

The Pyramid, the Louvre, views from outside during the day (Architect IM Pei)

DP Archive

The Pyramid, the Louvre, views from outside during the night (Architect IM Pei)

Inland Revenue Sports Centre, Nottingham

The translucent roof is poised above a glazed façade, so it might be considered a mixed form of a different type (a translucent form associated with a glazed form). At night the interior will glow with light both as seen through the glazed façade, and above it through the light escaping from the translucent roof.

Perhaps not so easily defined as the translucent roof, the open 'reflective' roof performs much the same service in defining the space below. This is a building form created for the provision of shelter, as for example bus or railway terminals. The inside of the roof clearly makes the major impression of the building seen from the exterior; for this reason, while it is clearly important that the interior lighting must be sufficient to provide necessary functional light to satisfy the needs of those using the space, this is insufficient on its own. The lighting performs a dual role; the

DP Archive

The Eden Project, interior (Architect Nick Grimshaw)

DP Archive

The Eden Project, exterior (Architect Nick Grimshaw)

DP Archive

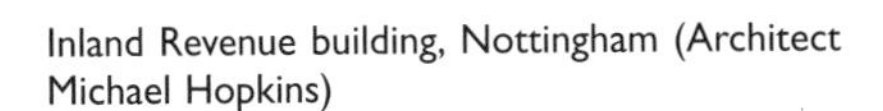

Inland Revenue building, Nottingham (Architect Michael Hopkins)

underside of the roof must be independently illuminated, in order to establish the space.

North Greenwich Transport Interchange

The North Greenwich transit building by Norman Foster is a good example of this style, conforming to the description of a reflective roof form. It is a building which covers not only a large bus terminal serving the Millennium Dome, it also houses the entrance to the Jubilee line tube station coming within its curtilage. As the building therefore has to cover a very large area, with many people converging both from the tube station and the buses serving the site, it has to establish a presence in the locality, as it is from beneath its roof that the public approach the dome, and from which they depart.

A special light fitting was developed between the architect and the manufacturer, to provide both upward light to the underside of the canopy, and downward light for the function of the building with the lit roof relying on the reflective properties of the roof material for its effect.

Peckham Square, London

A further example of the reflective roof form used to establish a spatial environment in a transport context. The roof is uplit with batteries of

Courtesy of Foster & Partners

North Greenwich Transport Interchange (Architect Foster & Partners; Lighting designer Claude Engle; Manufacturer l'Guzzini)

spotlights supported on special 'lighting trees' providing a dramatic appearance, the colour of which may be altered by means of filters.

Warner Village, Cribbs Causeway, Bristol

A final example of the use of a reflective ceiling as a canopy to a cinema and entertainment complex. The form of the canopy designed by the Engineer Ian Duncan makes a dramatic statement in the context of Cribbs Causeway, which is basically a large commercial park set amongst a sea of car parking.

Courtesy of Ron Haselden

Peckham Square, night view (Architect McAslan)

DP Archive

Warner Village, Cribbs Causeway (Engineer Ian Duncan)

Conclusion

This analysis of building form has attempted to show the relationship of form with the different methods of external and building lighting. The importance of buildings in defining the edges of exterior space in the urban nighttime environment, as they do during the day cannot be overemphasized, and for this reason the variety of ways in which they are experienced needs understanding as a principle determination of 'Nightscape'.

Part 2

3 Spaces between

Space around buildings . . . aesthetic . . . function . . . model studies . . . roads . . . pathways and cycle tracks . . . sport . . . parks, landscape and open areas . . . water and rivers . . . transport and parking

While buildings are generally the single most important element which will determine the perception of the nightscape of an area, the space between the buildings, in its various aspects, needs to be considered. Some areas, including parklands, are of considerable size; while others, such as pathways, take up only a small amount of space, but have an importance disproportionate to their size.

The analysis of 'Spaces between' is divided into eight areas as follows:

- Buildings: the immediate area surrounding the building
- Roads
- Leisure, pathways and cycle tracks
- Sport
- Pedestrian areas
- Parks, landscape and open areas
- River and water
- Transport

The various functions of the areas listed above are discussed individually; but there are one or two points having general applicability. These may be listed under two main areas: aesthetic and functional.

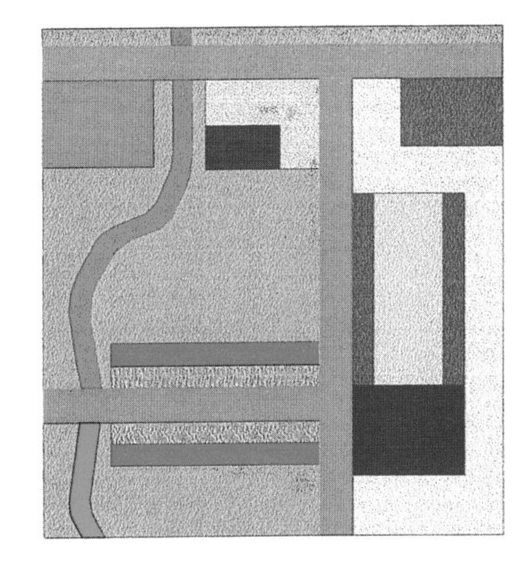

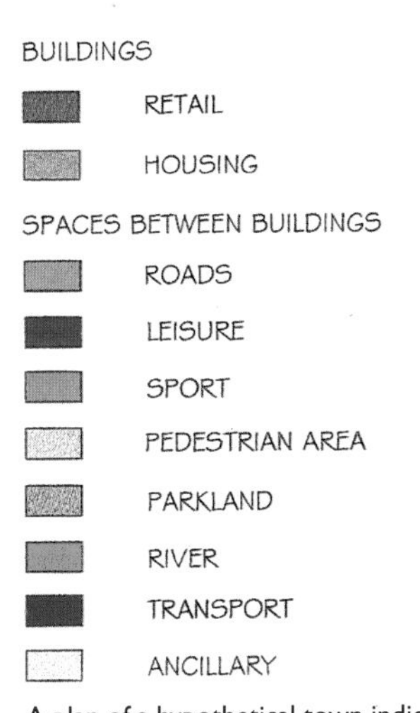

A plan of a hypothetical town indicating how the different areas might be disposed

AESTHETIC CONSIDERATIONS

Defining edges

Spaces are generally limited by some form of edge, and in the case of the urban space, this is more generally a building. Buildings have already been discussed, and the lighting of their façades form the most important aspect of the nightscape in defining the edges of a space. It will be seen from many of the city centre schemes illustrated later under the heading of Lit Environments (Chapter 6), that where the façades have been lit, there may be little need to add more than lit incidents to the spaces between, to achieve sufficient ambient light.

Where this is relied upon, it raises the question of whether the building owner is prepared to keep the lighting energized for as long as the space remains occupied, often late into the night. A particular example of this is the Place de Terreaux in Lyon (p. 74 and cover), where, with the façades of the building lit, there is sufficient additional light for the square from the main fountain, coupled with the pattern of lights set into the paving stones, which can be converted to lit fountains. At night the square is much used with its accompanying restaurants, open late into the night.

The defining edges are of course not always buildings, and where the space is parkland, this may equally well be a row of trees, but whatever their nature the lighting will need to be considered.

Glare

As with all lighting design the aim must be to eliminate glare. This is of particular importance where the need is for surveillance; but the human need is of greater importance, since too high a contrast between the different elements of an exterior environment will cause visual difficulty.

In identifying the problems it cannot be put better than in the excellent book *The Design of Lighting*.[1]

1. Is the lighting glaring to pedestrians or drivers?
2. Does it confuse visibility of traffic signs?
3. Does it shine into habitable rooms?
4. Is light shining into the sky reduced to a minimum?

Unity

The need for unity has been stressed as it applies to buildings in terms of the question of day and night appearance. It is equally important when applied to an open space. Unity is the particular province of the architect who is responsible for the space, either new or old. He will be ensuring the coordination of materials of the building, its scale and colour, and the way in which the lighting contributes to the overall effect may be built into a visual masterplan.

FUNCTION

Illuminance/energy

The question of 'Recommended Illuminance' for the widely different circumstances which prevail at night is dealt with in more detail in Chapter 5 (see Table 5.1, p. 88).

Illuminance levels will impact on the question of energy use, and decisions must be made in terms of the efficiency of the lamp, the hours at which the installation will be run, and the economies which can be made by an appropriate control system. In todays' low energy culture it is important that the type of lamp required should be chosen with care. Where the hours of use are likely to be low, it may well be that the less

---

[1]Tregenza, P. and Loe, D. (1998) *The Design of Lighting*. E & FN Spon.

efficient and less expensive lamps, such as tungsten halogen, may be appropriate; but where there are likely to be long hours of use, a calculation must be made between the higher initial cost of energy efficient lamps, and the savings involved taking into account their longer life, and lower use of energy. The question of cost in use is dealt with later (see in Chapter 5, Costs/energy) where the combined initial, capital cost, and running costs are considered.

Safety

There are two aspects to the question of safety: those of the safety of the equipment, the light fittings and the lamps to be used, and the safety of the lit environment itself, and those who use it.

Safety of the environment

The need to ensure that the lit environment presents no hazards, and that the lighting design enables a person to negotiate a path, or changes in level, with ease so that they can see where to go without difficulty. Any directional signs to be visible.

Closely related to the demands of safety, the needs of security require that a survey be made to ensure that there are no dark corners in which those with ill intent may lurk. This may need to be closely related to the requirements of CCTV cameras, to enable a watch to be kept at a distance. There is no doubt that a well-lit environment deters crime.

Safety of the equipment

Exterior lighting equipment often will, by its nature, be in close proximity to the general public. For this reason, unlike the interior lighting of buildings, which will be under the supervision and inspection of building staff, special care needs to be taken in the location and specification of exterior lighting equipment. The location, whilst being an aesthetic consideration (day/night appearance) is in addition a functional choice. The specification of the equipment must meet the rigorous needs of an exterior exposure, made worse in certain circumstances, such as marine conditions. More than this the chosen location must allow for periodic maintenance, lamp changing and accurate realignment.

It scarcely needs to be emphasized that where lighting equipment is placed in situations which are difficult to reach and therefore to change or clean, necessary maintenance is less likely to get done. Fittings placed at low level will need to be vandal resistant. It is important that the cost implications of a system of planned maintenance are taken into account at the outset when the overall costs of the scheme are being considered.

The importance of this has recently been recognized by the client of a high-profile floodlighting installation who contracted the original lighting design team to do annual maintenance and realignment to ensure that the original design intent is retained.

These are the functional aspects but the aesthetic aspects are of equal importance, and glare from inappropriately located or aligned lighting equipment can spoil the beauty of what might otherwise be an excellent floodlighting scheme, seen from a different direction.

Courtesy of Thorn Lighting

Goldsmiths University, London

(a)

(b)

(c)

(d)

(e)

Courtesy of Bartlett School

BUILDINGS AND THEIR IMMEDIATE SURROUND

There will be many situations where the overall environmental lighting, from buildings, street lamps or other sources, will make it unnecessary for special equipment to be provided. But in country areas for example there may be a need to provide some form of light guidance at strategic points around a building, and this may be done with wall-mounted equipment integrated into the building design, possibly with proximity switches providing light only when needed, or by pathway lighting at low level.

The nature and level of light will depend upon the building type and the area, it being important to ensure a colour relationship with the overall nightscape, and to fulfill the functional needs as already

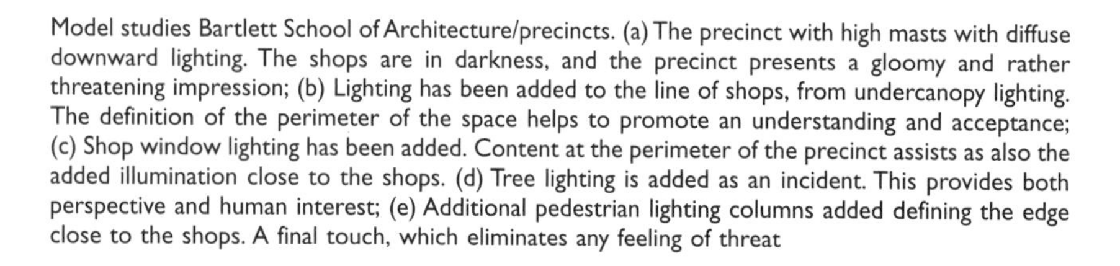
Model studies Bartlett School of Architecture/precincts. (a) The precinct with high masts with diffuse downward lighting. The shops are in darkness, and the precinct presents a gloomy and rather threatening impression; (b) Lighting has been added to the line of shops, from undercanopy lighting. The definition of the perimeter of the space helps to promote an understanding and acceptance; (c) Shop window lighting has been added. Content at the perimeter of the precinct assists as also the added illumination close to the shops. (d) Tree lighting is added as an incident. This provides both perspective and human interest; (e) Additional pedestrian lighting columns added defining the edge close to the shops. A final touch, which eliminates any feeling of threat

identified. In the Goldsmiths University illustration light is provided both from a regular spacing of uplights, recessed into the path, with added flower-bed lighting from an associated open area.

In 1989, model studies were carried out at the Bartlett School of Architecture and Planning, to establish some guiding principles for the lighting of the precinct of a shopping centre (presented at the CIBSE National Lighting Conference, Robinson College, Cambridge, in 1990).

The illustrations are of the model, and while we may have moved on from here, it is interesting that the principles set out as a result of the studies still have validity.

The purpose of the study was to indicate how different lighting methods can be built up to create a more humane atmosphere in what can be a very dreary space, so often covered as it is with car parking. The presentation concluded:

> Lighting for pedestrian areas must form a balance between good seeing conditions over the entire area and a light pattern that provides an attractive and welcoming visual environment

ROADS

Road lighting is a specialist subject and in the UK has been mainly the prerogative of the Institute of Lighting Engineers (ILE) composed mostly of Local Authority engineers, with assistance from the manufacturers of street lighting hardware.

Architects have, until recently, taken little interest in the design of road lighting, the design being left to the pressures of local authority economics. This often led to schemes which had poor colour rendering, were unrelated to functional street patterns and to glaring conditions due to wide column spacing.

Courtesy of Woodhouse

Oxford Street, London

Courtesy of Howard Brandston

Ottawa street lighting

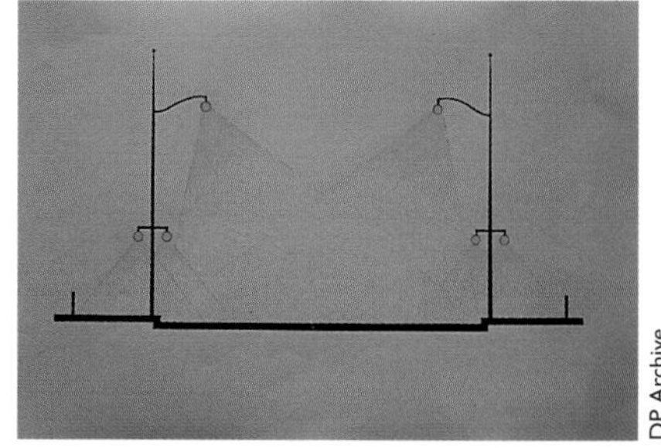

DP Archive

Sketch of road lighting

Happily this situation is changing: the use of cut-off lanterns, which direct their light downwards to minimize glare using lamps of acceptable colour, is more common. There is also acknowledgement of the different requirements of cars and pedestrians, as suggested by Waldram some time ago.

The illustration on page 53 shows a new road lighting scheme in Oxford St in London where the needs of cars, and those of the pedestrians are catered for by lanterns at different heights with different lamps from the same column, a device which is becoming more common. A similar system was used some years ago in Ottawa, as suggested in the above illustration.

Shop lighting

There is more to road lighting than lighting in the street. Roads are defined by their edges, which in towns will be the buildings, and often by shops. Due to the obvious hazard of combining shops with traffic, there is a move towards pedestrian shopping precincts, leading to covered shopping centres. These will have their own aesthetic, but in each case the purpose of the shop itself will be to attract the pedestrian, and there are many ways to accomplish this.

One way is to open up the shopfront with large areas of glass, allowing a panoramic view to the whole interior; the antithesis being the Chinese landscape approach – blocking the view in – allowing only a small open area to reveal a limited prospect of the goods on display.

There are many approaches in between, but the aim must always be the attraction of the goods on display. The result in the first case is a pavement which extends into the adjacent buildings giving light outward, while in the second there is a more subdued atmosphere, with little light extending beyond the shop, but which has its own attraction.

In both cases, some limited external lighting will be required. As in the shopping mall so evident in the old-fashioned high street with its

DP Archive

Intimate shopping street in Covent Garden

diversity, often having an appeal over the newer shopping mall which now tries to imitate it.

The introduction of a visual masterplan for our towns and cities would greatly assist in the field of road lighting, as in so many others, in that the function of roads could be identified by the nature and colour of the lamps used, the illuminance being associated with the nature of the road surface. One can envisage a directionality to the street lighting to assist car drivers to find their way, and to create a safer environment for the pedestrian.

LEISURE, PATHWAYS AND CYCLE TRACKS

Assuming that the pathway is unrelated to any structure, on which luminaires might be mounted, or roads with their own lighting, the first decision to be made is whether there is any need to provide lighting at all. The decision will be made on aspects of safety, and whether it is part of a public footpath system where the lighting may have to be maintained at least until midnight. Footpaths through parkland which is used by the public after dark should be provided with light either from smallscale columns or low-level bollards. These will be used by members of the general public who are night adapted, so that the level of light can be low; the danger which needs to be addressed here is where the path joins a more highly lit area or road, it may be necessary to provide for a gradual diminution of light from the brighter to the less bright pathway.

The illustration shows a cycle track, known as a 'redway' in the town of Milton Keynes, where small pedestrian scale columns spaced widely apart provide sufficient light for cyclists or pedestrians. Here there is no competition from more brightly lit areas, and therefore the level of light may be quite low. Where possible, the colour of the pathlight source should be correlated with the lamps used in other pedestrian areas.

DP Archive

Milton Keynes 'Redway'

SPORT

Areas set aside for sport activities vary from the open space in a village – usually the village green – to areas dedicated to specific sports, such as tennis, baseball, football, etc. which are likely be used both during the day and after darkness falls, and in some cases with large groups of spectators.

In some cases, for example on a golf course, there will be little need for nighttime lighting, although sometimes provided; whilst in other cases carefully designed sports lighting will be required.

It is important that whether the area set aside for sport is in the centre of a town, in a residential suburb, or in the countryside, the problems associated with light trespass or pollution are addressed.

The technology of the floodlighting design must be sufficient to ensure that light is provided only where it is required. The illustration below depicts an installation in a residential area where the light from the floods lighting the sports field, provides an unacceptable light to neighbouring properties.

In contrast, a successful example is illustrated by the lighting of a hockey pitch in suburbia lit to county standards; it is clear that the floodlighting design of the pitch has been carefully controlled to avoid spill-light beyond the pitch, thus avoiding light trespass to neighbouring properties.

Four 2 kW metal halide projectors mounted on each of eight 16 metre columns were provided (32 projectors in all) using cut-off lanterns with an asymmetric distribution. The illuminance level on the pitch was 300 Lux, providing county standards of play, but by simple controls it was possible to halve this for practice games, thus saving energy.

This was an installation carried out by a local authority, after a planning application for floodlighting the pitch had been called into question by local inhabitants, supported by their Member of Parliament. The residents were concerned about the question of light trespass. However, once the lighting of the pitch was installed no complaints were received, indicating the need for careful design, using well-engineered equipment designed for the purpose. It would also have saved time had the concept of public participation been applied.

DP Archive

Unacceptable light trespass

DP Archive

Teddington Hockey Field, acceptable lighting. Controlled floodlighting of a hockey pitch (Lighting designer DPA Lighting Consultants)

Courtesy of Thorn Lighting

Parc des Princes. Paris (Lighting Design Thorn Lighting)

The lighting of national sports grounds is clearly of a highly specialist nature evidenced by the Parc des Princes, Paris, illustration but the importance of light control is of no less importance in the small city recreation area used after dark to ensure that surrounding areas are not inflicted with unacceptable overspill. Information on the lighting for all types of sports is available in the CIBSE Lighting Guide LG4[2] which deals with the subject in detail.

PEDESTRIAN AREAS

Areas of towns set aside for pedestrians have always provided enjoyment, such as St Mark's Square in Venice, but areas such as these would

---

[2]CIBSE *Sports*, Lighting Guide LG4, 1990.

be categorized as whole lit environments. There are now, however, smaller areas of towns, associated in the main with some form of commercial development, which are set aside solely for the use of pedestrian traffic, except perhaps for a few hours in the early morning which are used for servicing vehicles at a time when the general public would not be present.

Like the parks and open spaces of a town these pedestrian precincts are very special areas, which should be designed at the human scale, providing safe access and security during the day, and an atmosphere of welcome, warmth and relaxation at night. High levels of light are not required, but any hazards, such as changes of level need to be well-lit. They may well provide for some landscaping and seating, and it is important that the lighting design assists in providing a sense of welcome, in addition to ensuring safety and security. An overall bland lighting scheme derived from lanterns on tall lighting columns, often found in municipal schemes, will not entice people to visit and revisit the area unless additional accent lighting is provided. An alternative method would be to install relatively low level lights giving emphasis or pools of light at functional points such as seating or landscaping, with which it may be associated; but it is important that dark, shadowed areas are avoided.

The problems of vandalism seems to be unknown in situations where such areas attract large numbers of people, as in Hong Kong or Holland. In Holland for example, very little additional light has to be provided in pedestrian shopping streets, where sufficient light is derived from the shops on either side.

The following illustration shows an American example of a pedestrian 'Transit' in Denver allowing access between shopping streets, cars not being allowed during normal hours. In this case a smallscale column with a decorative lantern has been designed, giving a warmth of light to the area without recourse to heritage fittings. The whole design benefits from the use of landscape to give it human scale. Likewise, the little pedestrian square in Doncaster.

DP Archive

Denver Transit, pedestrian area at night (Lighting designer Howard Brandston)

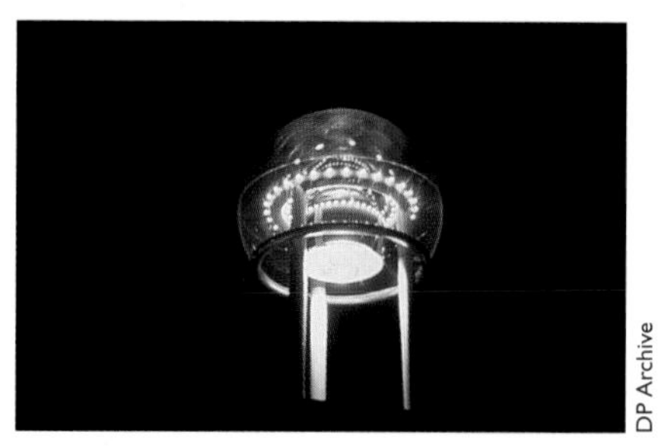

DP Archive

Denver Transit, detail of lantern (Lighting designer Howard Brandston)

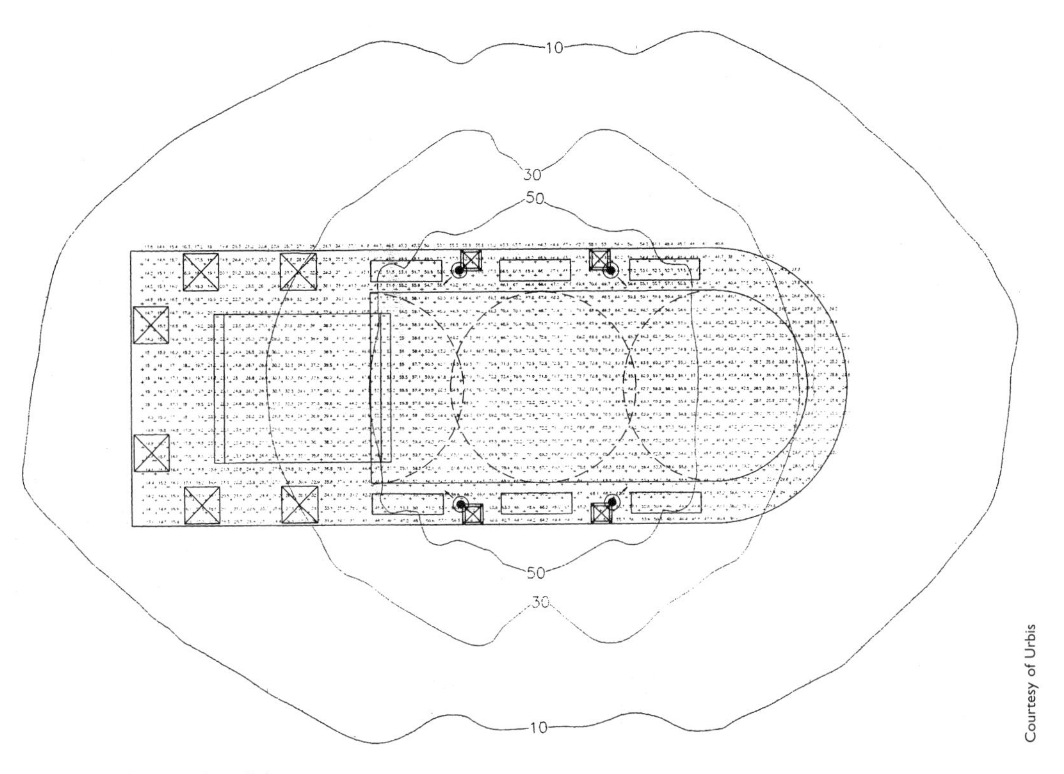

Courtesy of Urbis

Square in Doncaster, plan of the square

Courtesy of Urbis

Square in Doncaster by day (Lighting designer Urbis)

Courtesy of Urbis

Square in Doncaster by night (Lighting designer Urbis)

Parks, landscape and open areas

Described as the 'lungs of the city' the parks and other open areas in our towns are under continual threat from the forces of commercial development; it is only because London has its royal parks that it has been able to resist the force of change which would cover the city in concrete. New York has its Central Park, and Paris the Luxembourg Gardens and, for whatever reason, most great cities are well-endowed with landscape and open areas allowing the city to breathe.

No park is the same as any other and it is the infinite variety of environment, of layout and design changing with the seasons, which makes them a critical factor in any visual masterplan for the area.

Where arrangements are designed for good public access and imaginative lighting – as, for example, in the Tivoli Gardens which is mentioned in the Introduction and as a lit environment, there is no reason why the European climate should not allow the best use to be made of our parkland. The first obstacle to overcome is the idea that 'doing the job properly' will be expensive – it is 'not doing the job properly' that will be more expensive in the longterm.

An extensive survey of the park will be necessary to identify points of interest in the landscape, routes through to be identified and decisions made as to which areas should be emphasized. Not all areas of the park will require lighting. There should be areas of darkness where a person can look up and enjoy the sky, providing contrast with the surrounding more brightly lit areas. Landscape lighting is about perspective and contrast, and should be handled by a landscape architect well-versed in landscape design.

There are basically four areas of concern: the paths, the flower-beds, the trees and incidents. Each will need to be considered separately and together, so that the unity of the park is preserved.

Pathways have already been discussed above, flower beds are best dealt with by low level but flexible lighting which can react to the changes of the season, and tree lighting is best dealt with from below, which can be achieved by fittings recessed into the ground.

Tree lighting in a park in Hemel Hempstead shows the impression gained when different light sources, both cool and warm, are employed. It should be emphasized that landscape lighting should be dynamic, reacting to the changes in the colour of the leaves and type of tree. Where there is the advantage of water, this will enormously increase the impression of the lit environment by reflection.

Conversely, regarding the open countryside, the implication would be that there must be no nighttime lighting – and this would apply to the traditional village green in the UK, where it may be used for cricket or football during the day at different times of year, but perish the thought that it should be lit at night. There may be occasions where the green is used for a fairground having its own electricity supply for a limited period, adding a degree of variety to the scene.

Where an installation of nighttime lighting is appropriate, account must be taken of the importance of maintenance, and concealment of the light sources, and only equipment designed specifically for its purpose should be selected. There are many ways to conceal light sources amongst the landscape, and this is where the expertise of a landscape designer would be welcome (a useful reference for the lighting of parks and gardens is the CIBSE Lighting Guide LG6, pp. 24/25/58).[3]

[3]CIBSE *The Outdoor Environment*, Lighting Guide LG6, 1992.

DP Archive

Hemel Hempstead Water Gardens, view during the day (Lighting designer Dacorum Borough Council)

DP Archive

Hemel Hempstead Water Gardens, lit trees at night (Lighting designer Dacorum Borough Council)

DP Archive

Hemel Hempstead Water Gardens, trees and bridge (Lighting designer Dacorum Borough Council)

DP Archive

St. James Park, floodlit lake (Lighting designer Westminster Council)

River and water

Whether it is the light from an evening sunset seen reflected in the still water of a lake, or the underwater light source in a domestic swimming pool, the relationship between light and water has had an aesthetic appeal over the centuries, as much alive today as in the past.

The question of fountains is dealt with in Chapter 4 devoted to Incidents. Our concern here is with the spaces between created by streams, rivers and lakes, where the addition of artificial light sources with water may not always be appropriate, the evening sunset being sufficient.

In the illustration showing the Civil Aviation Authority offices, it is the light from the buildings seen in the lake which is sufficient to inform the nightscape, no additional artificial light sources being required.

Where a decision has been made to add lighting to a water scene – a decision which should be made with full recognition of the likely capital and running costs – the problems will fall into two categories: aesthetic and functional.

DP Archive

Village green in Hertfordshire, used for cricket in the summer

DP Archive

Village green in Hetfordshire, the arrival of the local fair

Aesthetic

A full brief needs to be worked out as to what appearance it is desired to achieve, and how this will change in rain or fine weather, the nature of the water (smooth or moving) and changes of time and season. As with all lighting installations, the question of the concealment of the sources to ensure the elimination of glare is of the greatest importance, perhaps even more so in that glare may occur both from the light source itself, and from its reflection in the water. Moving water will break up reflections, and if the water is permanently moving as, for example in a waterfall, the question of glare may be less of an issue.

The major decision here is the question of whether the light sources be placed below the water or above, raising the functional questions of electrical safety, security, and ease of maintenance.

The small floodlight placed below the water of a domestic swimming pool will make all the difference to the nature of the pool once night falls, its lit appearance making continued use a pleasure as well as safer. This need not be expensive, and is a small item in the overall budget for the pool itself.

With large expanses of water, and particularly where the natural water level may vary, the underwater option is less feasible, and Leeds Castle in Kent is lit from five projectors concealed around the verge of the lake above water level (see p. 20). This is a most economic installation, lighting as it does a major historic building, with visitors benefiting from the reflections of the Castle across the water.

Functional

It has already been stated that the functional problems associated with light and water are concerned with electrical safety, security and ease of maintenance; the most important of these being electrical safety. Wherever an electrical current at 240 volts (or in the USA 110 volts) is associated with water, special precautions need to be taken.

One method which overcomes the problem is to use remote source lighting, sometimes thought of as fibre optics, where the electrical power can be disassociated from the emission of the light. The light source itself

Courtesy of BDP

Civil Aviation Authority (Architect BDP)

can be placed in a location where it can be maintained easily and made secure, eliminating any electrical danger; from this point one or more light outlets can be placed at a distance from the power source, either above or below water, to produce the desired lit effect. A further example of underwater lighting is that used at the Hyatt Regency Hotel in Thessaloniki, where a rock wall is lit in such a way that the light sources are completely concealed below the turbulence of the water.

The other functional problems will no doubt be dealt with by the electrical engineer rather than the architect, but are no less important where they conflict with the architect's chosen locations; it is important that the elimination of glare is considered at all times.

DP Archive

Domestic swimming pool, using a single 300 watt LV halogen lamp

TRANSPORT AND PARKING

Areas contained in this category would include both exterior and covered parking, for public and private use, coach stations and garages. While each type will have its own particular requirements there are some general points that can be made. The first is to emphasize that in any project where cars or larger transport vehicles are involved, there will be a dangerous mix of pedestrian and vehicle, and steps must be taken to ensure maximum safety and security.

In the case of parking, whether open or covered, there is potential danger, and steps should be taken to avoid the appearance of a threatening atmosphere. It is not coincidence that film violence often takes place in multi-storey car parks – where there is unsupervised and poor lighting.

Entrances and exits to such areas should always be well-identified together with signage and directions. Once inside the parking area, the location of parking bays should be well-illuminated and where there are pay machines these should have special emphasis. It has been well-established that well-designed and adequate lighting deters bad behaviour and vandalism, and well-lit car parking encourages public use. It will repay the investment required.

DP Archive

North Greenwich Transport Interchange, day view (Architect Foster & Partners; Manufacturer l'Guzzini)

Covered transport buildings are exemplified by the North Greenwich Transport Interchange (also see Chapter 2) where there is combined use of an underground station connected with a bus station; this has a lighting system which emphasizes both the nature of the space and provides good illumination for the pedestrians who use it.

Courtesy of DW Windsor

Bluewater car parking (Lighting Designer Speirs & Major)

In this example special luminaires were designed to light upwards to a highly reflectant ceiling, to establish the nature of the enclosed space and downwards to the public area below. The building is used as a covered bus waiting area, for a series of buses which stop outside, the public being protected from the weather when necessary. In the past bus stations have tended to be rather formidable areas where there was a danger to the public from the buses which moved in or out. It may not always be possible to provide much space for vehicles to arrive, and where buses have to stop next to each other, the space between being used for boarding, needs to be particularly well-lit if accidents are to be avoided.

Petrol stations are a further category which have their own special requirements where the first impression, that derived from an approaching car, needs to be one of welcome and ease of access. The pump area must be immediately recognizable and the self-serve pumps easy to operate, with visual access to the payment point, so often associated with a small retail outlet. The illustrations depict the difference between a station relying on downward light to the pump area and one where the canopy is up-lit from the pump position, ensuring that the canopy is well-lit, and the pump area clearly visible.

DP Archive

DP Archive

Comparisons of typical petrol stations. Top, upward lighting; bottom, downward lighting

CONCLUSION

This chapter has attempted to explore in very general terms the spaces between buildings, but the author is conscious that it has barely touched on some aspects, such as road lighting, or sports, where the specialist engineering problems need to be considered in greater depth. At best, it has indicated that it is important that the aesthetic and functional problems posed by the spaces between should be regarded as important environmental issues.

4 Incidents

Walls . . . gateways . . . bridges . . . fountains . . . statues/sculpture . . . trees . . . ephemera

In the preamble to the previous chapter on Spaces between, the aesthetic and functional considerations which should inform discussions before a proposal for a night lighting scheme is designed, can be applied equally to those aspects discussed under the heading of Incidents.

'Incidents' are defined in the Oxford English Dictionary as 'events of accessory or subordinate nature' and since it is difficult to find a more appropriate word to describe the statue in the square or the fountain in the park, the term is extended to include other events of a grander or, in some cases, a more ephemeral nature.

The following incidents are included:

1. Walls
2. Gateways
3. Bridges
4. Fountains
5. Statues/sculptures
6. Trees
7. Ephemera

and each type of incident is discussed below.

No doubt an exhaustive search would reveal other 'incidents'; but it is believed that the list covers the majority and that where the list is deficient, the examples given will illustrate the principles which may then be applied to other areas.

WALLS

It is important to establish at the outset the purpose for which the wall is to be lit, whether this is a background or whether the wall itself is of significance due to its sculptural quality. The lighting of a wall which may define a space, is perhaps the most simple of lit incidents, but it is often the case that the geometry of the lighting is poorly designed, leading to uneven light at the base of the wall, or lighting which spills above the wall, leading to light pollution. In some circumstances, other aspects which may need to be addressed are the ecological dangers which might exist.

In the case of the first example, the Chalk wall at the Bluewater Shopping Centre in Kent; this is seen first as a background to the buildings and car park areas, and a defining symbol of the colour chosen to represent the name and the site.

The method of floodlighting the wall is by means of low level floods filtered to provide the blue light to the chalk face. These are trained upwards from a sufficient distance from the base of the wall, allowing the light to reach far enough up the wall surface to give the impression that the whole cliff is lit. Further use is made for advertising purposes, by means of a gobo projecting the name of the centre onto the cliff side. If this is used for advertising it is an expensive method, as even in this case where there is a light coloured wallface, it still requires a significant expenditure of energy.

Courtesy of Spiers and Major

Blue Wall at Bluewater Shopping Centre (Lighting designer Speirs & Major)

A number of mountain surfaces have been emphasized by lighting and there is no doubt that they provide an impressive appearance; best when confined to a specific area. For example, at Mount Rushmore in the USA, where the mountain has been carved with images of past presidents, or at the Hyatt Regency Hotel, Thessaloniki where the wall is associated with water, which, due to its turbulence, conceals the floodlights placed below the water at the base of the wall (see also Chapter 6).

A further example of a lit wall is a carved war memorial at Tournon in France. The memorial and a small patio area in front are bordered by a ring road, separating the memorial from the normal public view; this allows the use of close offset floods recessed into the paving in front of the wall. From across the road the floods are completely concealed, and are designed to provide sufficient modelling to the sculptured form of the memorial. Here no attempt has been made to use saturated colour, allowing the warm colour of the lamp to bring out the natural colour of the stonework.

A sophisticated example of wall lighting is the wall below the pyramid at the top of Canary Wharf Tower. This is seen as a white band in the picture, but the fluorescent lighting installation is integrated with the design of the wall at the side of the plant room for the building, and is

Courtesy of Robert Honeywill

Hyatt Regency Hotel, Thessaloniki (Lighting designer Maurice Brill)

therefore protected from the elements whilst maintenance can be done under cover. The lighting installation is controlled by the BEMS for Canary Wharf, and is capable of colour variation to suit different moods and festivities.

GATES

Modern towns no longer have the traditional gateways, associated with fortified cities, and our new towns lack physical gateways or points of arrival, which can be emphasized by architectural expression during the day, and therefore lack the possibility of lit emphasis at night. Some

DP Archive

War Memorial in Tournon, France, the lighting method used

DP Archive

War Memorial in Tournon, France, the war memorial at night

Canary Wharf, the Tower (Lighting designer DPA Lighting Consultants)

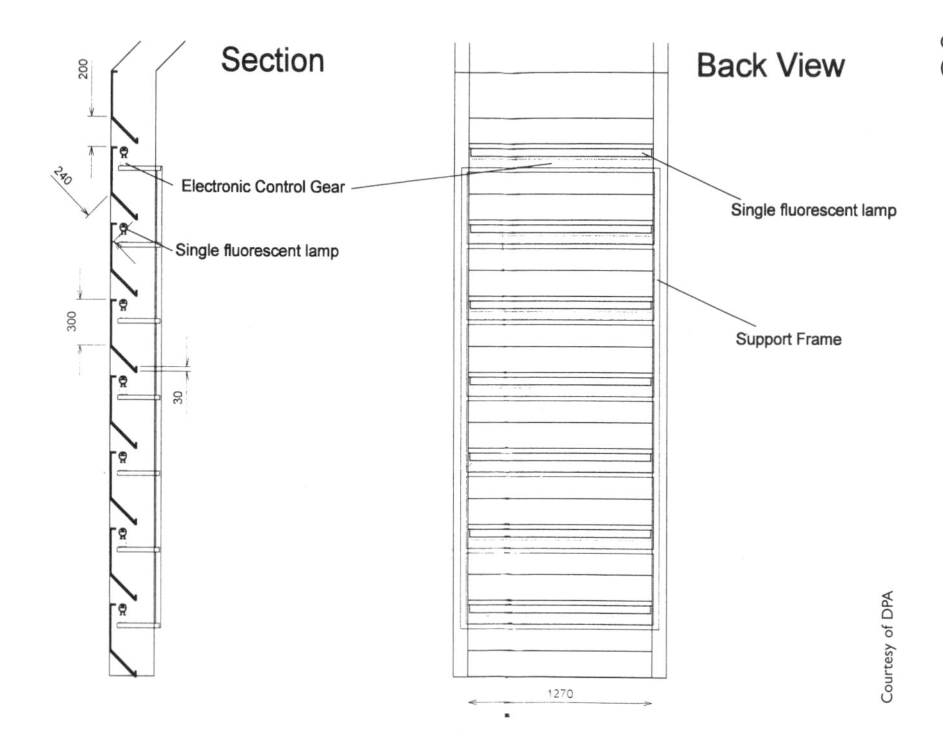

Canary Wharf, detail of section through wall (Lighting designer DPA Lighting Consultants)

efforts have been made to overcome this in strategies developed by lighting designers, as for example in Edinburgh, but lighting design alone cannot achieve this.

Gateways have always attracted attention as requiring some symbolic emphasis, and this has generally been accomplished by forms of decorative lanterns placed on the side posts of the gate, the light from the lantern spreading sufficiently to light the entrance pathway itself. The danger with this method is that there will be glare from the lantern seen against a dark background, and that there will be insufficient light at ground level.

The Queen Elizabeth Gate to Hyde Park is no exception, in that decorative lanterns have been placed on the top of the gateposts, but the problem of glare has been overcome by lighting the posts themselves by means of a row of recessed uplights at each post position. In addition, bollards with asymmetric reflectors placed on each side of the gate ensure that the intricate filligree metalwork registers; so what at first sight might appear to be a simple solution to lighting a gate, is in fact a sophisticated lighting design bringing out the beauty of the gateway.

At the other end of the scale is the lighting of the normal domestic entrance, the front door to a house. Unfortunately, the standard solution appears to be to use some form of coach light designed to throw light into the eyes of anyone approaching the door – the standard glare situation. When choosing a lighting solution for the front door it helps to clarify what you are trying to achieve, for example light to the door itself, to enable the key to be placed in the lock, light to the overall environment of the entrance to provide a warmth of welcome to arrivals, and light to the path leading up to the entrance for safety. There may well be other considerations such as light for security and CCTV.

BRIDGES

One of the most important incidents in a town is the bridge. It is unnecessary to cover the many different types of bridge here, but whether it is a small pedestrian bridge, or a main traffic artery, the way a bridge appears at night is of vital and often symbolic importance to the overall character and quality of the nightscape. A seminal example of this must be the Sydney Harbour Bridge which has, since it was built, been the 'image' of this city in Australia; now even more so, associated with the new Opera House.

The starting point must be an understanding of the bridge structure. The lighting must bring out its essential character, and to do this the light sources must, as far as possible, be concealed from view; it being the elements of the structure that should be revealed: the suspension of a suspension bridge; the solidity of supporting towers; the lightness of some designs; the heavy masonry of others, etc. Bridges have an important civic duty to perform when viewed from a distance and when at close quarters, a duty not to cause glare to others from the projectors necessary for the lighting design.

A possible exception to the rule that light sources should always be concealed, might be said to be the use of strings of small unshielded light Bulbs (or Zenon capsule lamps) used to define the lines of construction of a bridge. This has generally been ruled out by the transient nature of the lamps used, leading after a short while to lamp failures and gaps, but it is always conceivable that lamp technology may have been developed to

Courtesy of Nigel Pollard

Queen Elizabeth Gate, Hyde Park, lit gate at night (Lighting Designer Nigel Pollard)

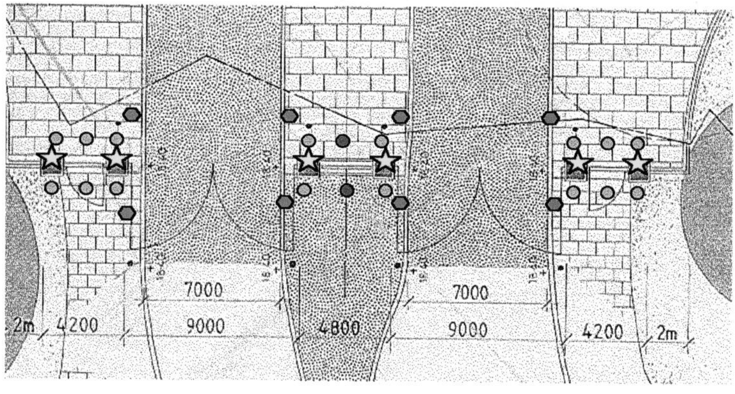

- 6no. Stone Pillars each lit from asymmetric ground lights with 6,000 lumen (70w) 3K metal halide lamps. *(Hydrel)*
- ☆ Each pillar topped by traditional gas lights. *(Sugg)*
- Central screen lit from single asymmetric ground light with 11,000 lumen (150w) 4K metal halide lamp. *(Hydrel)*
- Footway gates lit from single asymmetric ground light with 6,000 lumen (70w) 3K metal halide lamp. *(Hydrel)*
- Roadway gates lit from custom made bollards each containing asymmetric reflector system with 6,000 lumen (70w) 3K metal halide lamp. *(DW Windsor Ltd)*

Courtesy of Nigel Pollard

Queen Elizabeth Gate, Hyde Park, plan of lighting (Lighting Designer Nigel Pollard)

overcome this and its consequent maintenance problems; but it is doubtful whether this presents a longterm solution to the problems of lighting modern bridges.

The Erasmus Bridge in the city of Rotterdam is an example of these principles, in that the lighting designer persuaded the architects to place white sleeves over the suspension cables to reflect the light. The

DP Archive

The Albert Bridge, London (Lighting Designer Philips Lighting)

suspension cables themselves are lit from low level concealed projectors, together with the supporting towers; while the roadway across the bridge is lit with cut-off lanterns defining the road, without causing glare to the general view. It appears to be a simple solution, but required sophisticated lighting engineering in achieving its design – that of a vast sculpture in the sky, symbolizing the grain of the city.

Similar in essence is the lighting of the Normandie Bridge over the River Seine at Le Havre in France, and the Benjamin Franklin Bridge in Philadelphia. Both bridges emphasize the suspension cable structure: in the one case suspended at angles from the tall towers, and in the other, vertical cables suspended from catenary suspensions above. Each solution requires that the roadway itself is independently lit.

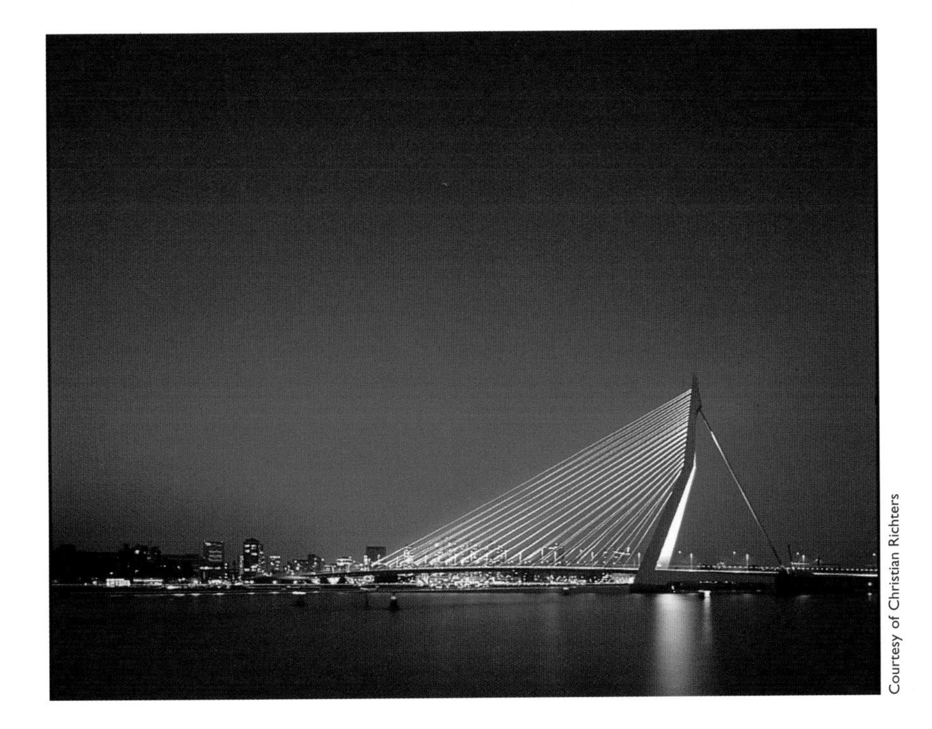
Courtesy of Christian Richters

Erasmus Bridge, Rotterdam, view of bridge from a distance (Lighting designer Lighting Design Partnership)

Courtesy of Christian Richters

Detail of lighting method (Lighting designer Lighting Design Partnership)

In the case of the Benjamin Franklin Bridge, an innovation is the sophisticated lighting control system that allows the lights to be switched rapidly on and off in sequence, providing a dynamic effect which relates to the frequent passage of commuter trains across the bridge.

Whilst suspension bridges may be the high flyers due to the possibility of bridging huge spans, and creating large civic statements, there are many other types of bridges, some of which require an alternative technology, where the size and construction will vary.

Courtesy of Matt Wargo

Benjamin Franklin Bridge, Philadelphia, general view of the bridge (Architect Venturi, Scott-Brown

Courtesy of Matt Wargo

Benjamin Franklin Bridge, Philadelphia, details of light fittings (Architect Venturi, Scott-Brown)

The Marshall Bridge in Berlin, only 28 metres long, is formed of simple reinforced concrete arch construction, requiring no overhead support. The method of lighting has been developed using four 8 metre circular towers, two at each end of the bridge. Each tower houses four projectors trained upwards towards mirrored reflectors designed to distribute their light downwards towards specific areas of the bridge, the 16 projectors being capable of covering the entire bridge area. No light or emphasis is given to the reinforced concrete bridge structure itself.

The antithisis of the Marshall bridge is illustrated by the bridge at Ville d'Alençon, in France. Here the heavy masonry of the bridge construction is emphasized at the expense of the roadway above. A modest scheme of light sources, concealed from the view of the public, light the underside of the arch, the arch appearing twice its actual size due to the reflection of the underside of the arch in the river. The lighting design is wholly appropriate to the structural character of the bridge (see Chapter 6, Lit Environment, p. 164).

The lighting of bridges will always offer a challenge, as the methods of structure change – from the Maillart bridges in Switzerland in the 1930s to those of Calatrava today – bridges are in constant development. The problems arise when the lighting conflicts with the structure, or fails to inform it; and sadly there are many well-known examples of this. The

Courtesy of Sill Lighting

Marshall Bridge, Berlin, distance view of the bridge (Lighting designer Sill Lighting)

projectors required for lighting large bridges tend by their nature to be bulky, and it is important that their positioning is both unobtrusive during the day, and does not provide unacceptable glare seen from positions close to the bridge at night; sometimes a conflict with the needs for the equipment to have good access for maintenance (see Parks, landscape and open areas, Chapter 3, p. 60).

Courtesy of Sill Lighting

Marshall Bridge, Berlin, details of the column light (Lighting designer Sill Lighting)

FOUNTAINS

As an incident in a town, the fountain with its moving water is unsurpassed, and it is no coincidence that in the Mediterranean climate of Italy, where serious frosts are uncommon, the architecture of fountains achieved its height. Therefore, if a fountain remains unlit at night it is a wasted opportunity. It cannot be beyond the wit of man to ensure that fountains can flourish in the cooler climates of Europe or the States – the technology exists nowadays to ensure that the fountains will not freeze up in cold weather – and it is not the technology which is lacking, but often the will to do it.

One of the world's great examples is the Trevi fountain in Rome illustrated in Phillips.[1] This is typical of the sculpture in a basin, where the sculptures rise from the water at low level and the light sources are concealed from the side, or from below.

The art of fountain design is full of innovation, and the different types of fountain are limitless. From the traditional sculpture in a basin to waterfalls or vertical jets set into paving, the lighting design also needs to be innovative, to suit the new designs. New lighting methods will be found as the technology develops.

As always, when light sources are associated with water, electrical safety is of the utmost importance, and there are good reasons for using low voltage sources, or remote source lighting. But of equal importance is

[1]Phillips, D. (1997) *Lighting Historic* Buildings, p. 175. Architectural Press/McGraw-Hill.

Courtesy of International Lighting Review

Bartoldi Statue, Lyon, Place de Terreaux, general view of fountain at night (Lighting designer Philips Lighting)

DP Archive

Bartoldi Statue, Lyon, Place de Terreaux, details of fittings concealed by rocks (Lighting designer Philips Lighting)

the manner in which the light is directed to the water movement, to enhance its sparkle and excitement. Ideally the source of light should be projected within the water jet itself.

There are basically two types of underwater luminaire. In the first case there are sealed cavities with suitable dry access, into which mains or low voltage fittings can be placed. In the second, low voltage submersible fittings can be located within the fountain basin and concealed by rocks or other features. In the latter case there would be considerable advantages in using fibre optics or remote source technology, and much equipment is now available for this purpose. A fine example of the traditional sculpture fountain is located in the Place de Terreaux in Lyon. The large sculpture by Bartoldi with its attendant horses sits in the water basin, and the light sources are placed so as to project light up to the elements of the sculpture, the sources themselves being concealed by artificial rocks within the water basin.

Of a similar type and period as the fountain in the Place de Terreaux, the fountain in the Place de la Concorde in Paris is, in every sense, a fountain rather than a sculpture, since it relies for its effect on the jets of water, rather than a central figure. The lighting relates purely to the water jets, and it produces its effect from the dancing light reflected off the patterns it creates.

An entirely different effect can be seen in the series of fountain jets in San Jose in California. Here, vertical water jets are set out in a symmetrical pattern leading up to a series of formal steps. Each jet is lit by a narrow beam submersible floodlight using a low voltage lamp set into the paving below the jet. The appearance, whilst being quite different

DP Archive

Fountain, Place de la Concorde, Paris (Lighting designer Philips Lighting)

DP Archive

Fountains in San Jose, California (Lighting designer Philips Lighting)

to the traditional pattern, has classical appeal, and there are examples of this type of fountain where the jets can be controlled to react to the sound of music, and change colour for special effects.

But fountains do not have to be major operations, and the little fountains in the gardens at Tivoli is one such example, representing the charm that can be produced from very simple means. Designed over 50 years ago, these little wooden flower bowls set into a formal garden, have a miniature low voltage lamp set below the water jet so that at night it glows with light (see also Tivoli Gardens, Chapter 6, p. 148).

Some of the simplest fountains are those that rely on the spread or flow of water over large flat stones. There will be a central orifice through which the water flows onto the top of the stone, but this will be too small for the introduction of all but the smallest light source, and other methods have to be found to provide the shimmer of light across the width of the stone. This may best be left to moonlight or whatever environmental light is available. The small fountain in a suburban garden in Hertfordshire shows the renewed richness of fountain design, with water gently spilling over the edge of the top stone to the lower basin, illustrating their timeless appeal. The lighting of fountains is not the only environmental impression that is made; it is also the sound of water which has impact. Fountains create their own magic in the environment, and the world would be a poorer place without them.

DP Archive

Tivoli miniature garden fountain during the day (Lighting designer Louis Poulson)

STATUES/SCULPTURE

If we ignore the fountain sculpture, as having already been discussed, there are two main types of sculpture or statue.

1. The traditional stone or bronze statue
2. The self-lit sculpture

The lighting of these two generic types requires an entirely different approach. Up to the twentieth century, there would have been only one way: daylight during the day and floodlight cast from a distance onto the solid form by night. But with the development of modern light sources, an entirely new concept has been achieved, where the light sources become an integral part of the structure.

The traditional stone or bronze statue

These vary from the small statue represented by 'The Little Dancer', in London's Covent Garden, to the Statue of Liberty in New York the one requiring minimal solution, and the other a major lighting installation; but both having the same basic criteria: to model the form of the statue, and to achieve this in such a manner that the light sources are unobtrusive if not invisible.

The starting point, as always, is a study of the nature of the statue by daylight, since in most cases it will have been conceived in daylight in the sculptor's studio but it must be accepted that the appearance of the statue at night will not be the same – indeed should not be the same – since efforts to reproduce the daylight experience would be ill-advised, and doomed to fail.

There will be a difference of colour, modern light sources being either cool or warm, or having saturated hues; but the biggest difference will be

DP Archive

Domestic fountain in a garden in Hertfordshire

the result of the direction of the light sources, and the modelling of the form these produce. Unless there are possibilities to locate light sources at high level from neighbouring buildings or other points, the light will generally be from below, creating the opposite effect to that of daylight.

An important consideration when making decisions on the location of the lights is the likely viewing points of the statue. Where this can be seen in the round it will prove the more difficult, but by far the more usual situation: there may be occasions where a more theatrical approach can be adopted, with a single viewpoint allowing a unidirectional solution, as when a sculpture is set against a wall. The 'bas-relief' is the epitome of the unidirectional approach, where light can only be thrown from a frontal position.

Since it will be difficult for a lighting designer to make decisions on the way in which the sculptor would like to see his work at night lighting trials should take place. These would be attended where possible by the originator of the piece, to enable the designer to demonstrate the lit form which can be produced, bearing in mind the possible locations and colour of the recommended light sources.

In the case of 'The Little Dancer' a convenient street lighting column enabled the designer to place a single metal halide (CDM-T) spotlight

Courtesy of Thorn

'The Little Dancer', Covent Garden (Lighting designer Nigel Pollard, NEP)

Courtesy of Pinniger and Partners

Statue of Rebellion, Trafalgar Square, depicting upward floodlighting (Lighting designer Pinniger and Partners)

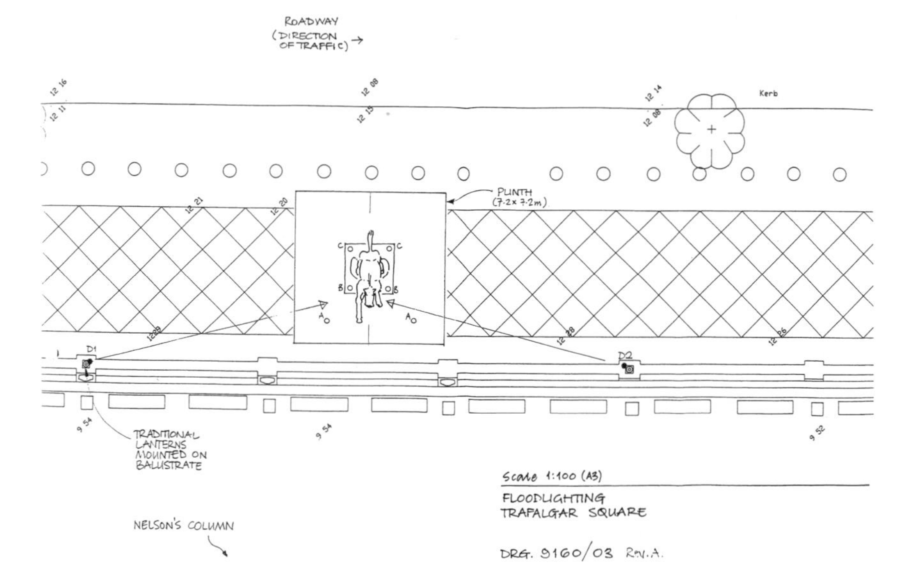

Diagram of light plan for Statue of Rebellion, Trafalgar Square (Lighting designer Pinniger and Partners)

7 metres above the statue; where, because of the narrow angle of the beam, there is no glare from the fitting, which throws an acceptable pool of light on to the front of the statue, the normal angle of view – an example of economy of means.

The Rebellion statue in Trafalgar Square is a further example of the solid form traditional sculpture, but here the more usual solution of lighting from below has been adopted. The statue on its stepped podium enabled an uplighter to be recessed in one of the higher steps, at a level which avoids glare to those walking in the square. The sculptor, Judy Boyd, agreed the important public view of the statue was by pedestrians in the square, so the nighttime appearance should emphasize the dynamic quality of the horse as seen from below, and this it clearly does.

In some circumstances, lighting from below can create an ethereal effect, as evidenced by the sculpture by M. Blick which, lit from below, the form of the sculpture appears to float, with little or no connection to its base.

DP Archive

Sculpture by M. Blick (Lighting Thorn Lighting)

The 'Angel of the North' by Anthony Gormley placed on a hill overlooking a motorway is a good example of a unidirectional viewpoint, so that a system of frontal lighting is all that would be required. Furthermore, since the sculpture is placed at high level well away from the public eye, there is no chance of glare from light sources placed at ground level. However, a decision has been made that at any rate for the present, the sculpture will remain 'unlit' at night – to be experienced by daylight during the day and seen in silhouette, by the moon and the stars at night. This is a perfectly valid decision. The tall all-metal sculpture can be seen from a considerable distance, and although only in position for a few years it has already become an important monument for the travelling public, by day, if less so by night.

A more complex example of the lit sculpture, is the Albert Memorial in Kensington Gardens, which has recently been renovated, with new lighting, and this is considered later in the final chapter of the book dealing with Lit Environments (Chapter 6, p. 114).

The self-lit sculpture

This is an area which is developing, so that there are comparatively few examples from which to draw on. However, one or two points can be made. The form of such sculptures is divided into those which exploit the

Courtesy of Arup Associates

The Angel of the North, by daylight (Sculpture Anthony Gormley)

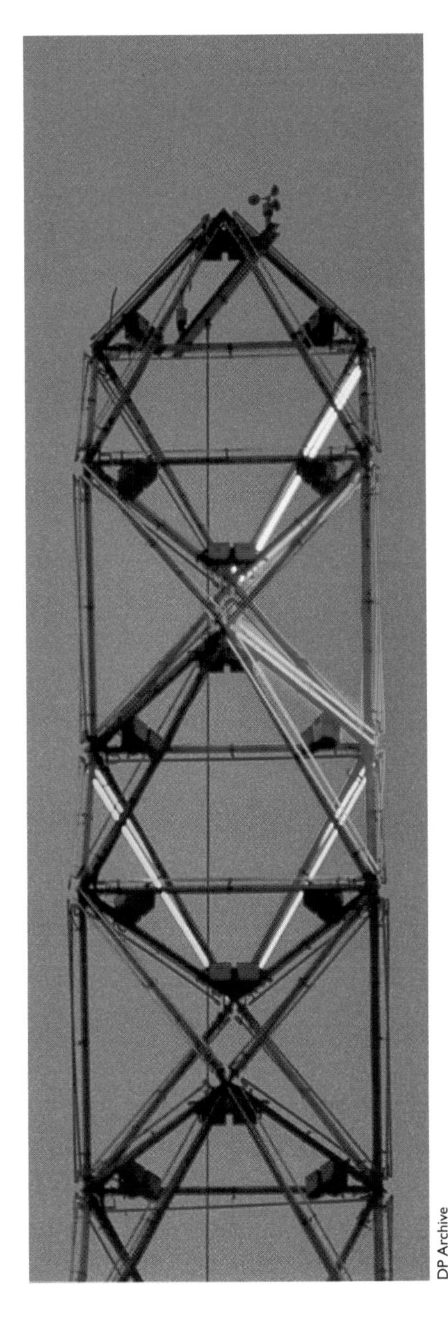
DP Archive

DP Archive

DP Archive

Fluorescent sculpture, Hayward Gallery

DP Archive

nature of certain lamps, and those which integrate the chosen light source within a design framework, the lamps themselves being concealed.

An example of the first type is the fluorescent sculpture mounted on the roof of the Haywood Gallery in the 1950s. Here the particular quality and colours available of fluorescent sources are optimized, together with a sequential change to make a dynamic visual experience. The colour changes are related to atmospheric data, wind speed, temperature and pressure to ensure that there is a dynamic visual pattern. Positioned on the top of the Hayward Gallery it immediately suggests the raison d'être for the building below, echoing 'change with the new'.

A newer example of this type is the Coquille – the egg designed for a roundabout in the city of Rheims. The stainless steel structure, completed in 1989 to celebrate the bicentenary of the French Revolution, recalls the shape of the city in Roman times. The sculpture is formed by two

Courtesy of International Lighting Review

La Coquille, Rheims, by artist Alain Le Boucher (Lighting designer Philips Lighting)

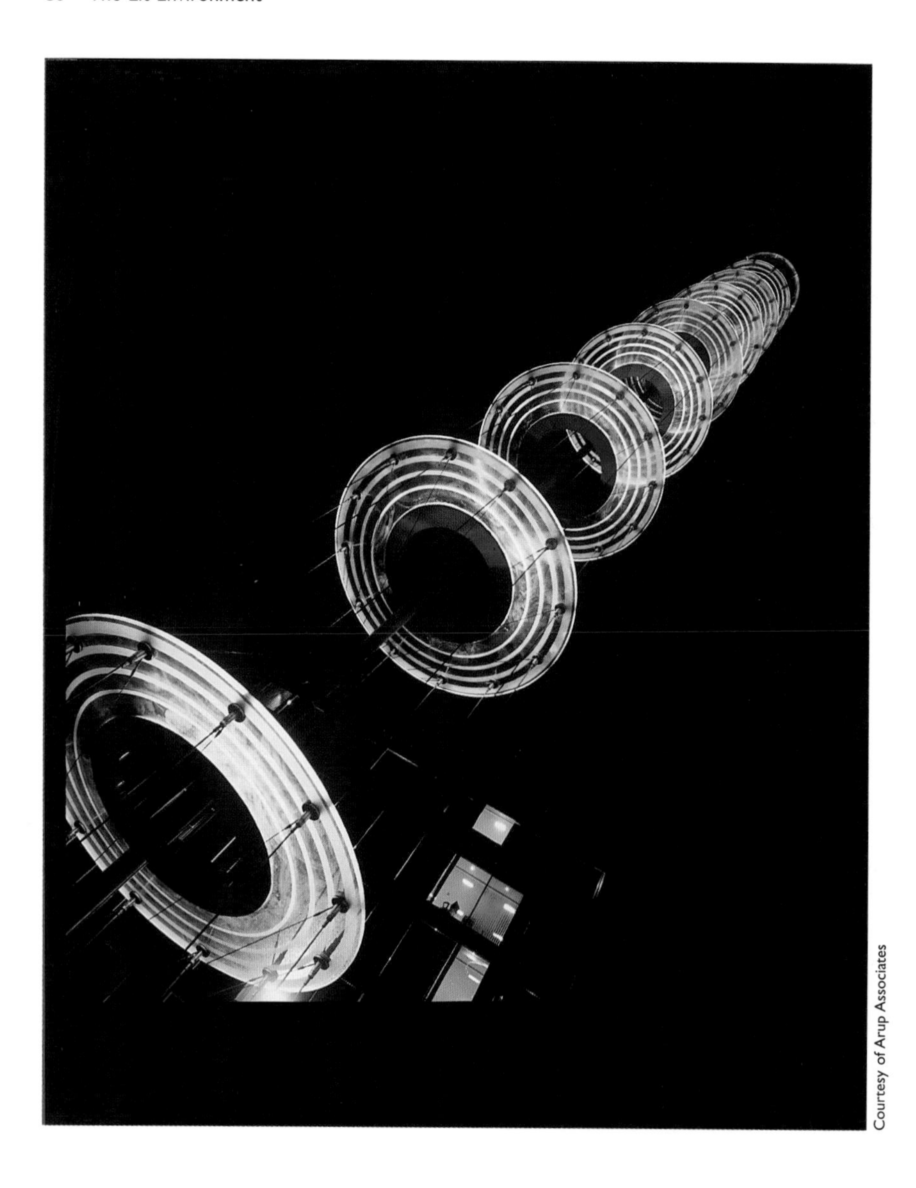
Courtesy of Arup Associates

Fenchurch St., Beacon, at night (Lighting designer David Hymas & Haico Sheppers)

interlacing grids, with lamps at each nodal point. At night the grids are seen in silhouette with the dancing light from the lamps controlled by computer to give a constantly changing light show.

An example of the second type is a glass and light feature in Fenchurch St, London. The Beacon, as it is called, comprises a lightweight stainless steel structure 27 metres high. The structure consists of a central pole surrounded by lightweight stainless steel strands, supporting and supported by eight 2-metre diameter laminated glass discs. At night the structure almost disappears, whilst the glass discs glow with light; light sources comprising a series of 18 watt compact fluorescent lamps contained in a circular luminaire box. Here the light sources are concealed, allowing their light out to edge-light the glass discs. The impression at night is of an illuminated column.

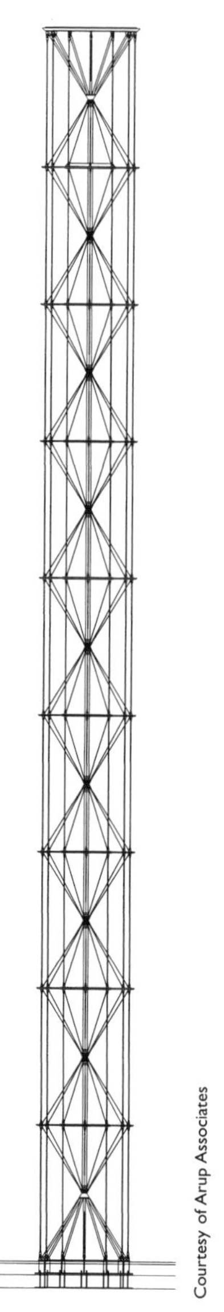
Courtesy of Arup Associates

Fenchurch St., Beacon, detail of construction (Lighting designer David Hymas & Haico Sheppers)

TREES

When planning the lighting for trees, which may at the time be quite small, or not yet planted, it is important to know the species, and the mature shape and whether they are evergreens. Trees which shed their leaves in autumn, are of a different nature, and it is important that the lighting should reflect the colour; if a single light source is planned it must be accepted that the appearance will change with the seasons.

The choice of light source is clearly important as to whether it is desired to reflect a natural colour, or whether a warm or saturated source is more appropriate; the most important aspect of all is the nature and location of the source. It has become almost standard practice to recess light sources into the ground below the tree, and there are a large number of well-designed and engineered light fittings which have been designed specifically for this purpose.

Courtesy of Building Design, Partnership BDP

Clarendon Dock, Belfast. Tree-lined avenue (Lighting designer Barrie Wilde)

An aspect which should be borne in mind, is that decidous trees cast their leaves in winter, so that whilst the tree canopy will accept the light during the summer, in the winter if the same light is used it will travel upwards to the sky through the bare branches; for this reason, it may pay to provide a change of light in the winter, or alternatively provide two systems, if light pollution is to be avoided. The effect of light on the winter scene can provide rewarding results by reducing the amount of upward light, thus reducing the impact of light pollution.

The trees at Clarendon Dock in Belfast are an example of this form of lighting where the fittings seen below the trees are comparatively glare-free, and the effect of a lit avenue of trees works well.

An alternative approach has been adopted in the Hemel Hempstead Water Gardens, already illustrated (see Chapter 3, p. 60) where fittings have been placed at ground level shielded by simple timber shields; these fit into the appearance of the undergrowth surrounding the trees. This example also shows the effect of the different colours of light source, warm and cool, that can be used to provide a variety of effects.

The sculptural effect of specific species of tree can be used, set against an illuminated background, to silhouette the outline of the tree which can produce a dramatic effect, an effect which will change with the seasons. The frontal lighting of a group of trees seen at a distance from a single direction, is often used by landscape architects to provide perspective to a view. The lighting of trees in the landscape may be considered as a whole environment, but the lighting of a single tree may also be considered as an important incident. A useful section on the lighting of different types of tree is contained in CIBSE[2] and a good reference for garden lighting is Janet Lennox Moyer's *The Landscape Lighting Book*.[3]

EPHEMERA

These are the lighting effects of a moment which is most perfectly illustrated by fireworks, here the ephemera lasts a few seconds, but for that moment produces an unsurpassed magic.

Ephemera or 'existing only for a day or a few days, short lived, transitory' (*Shorter Oxford English Dictionary*, published by Oxford University Press) encompasses many different light forms, and would include festive lighting such as Christmas lights in a shopping street, the lighting of a building for some special occasion, *son et lumière* and many

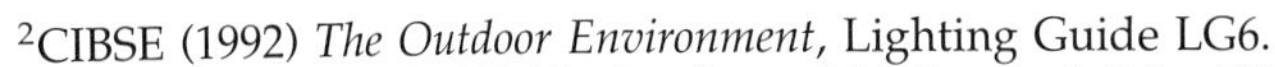

[2]CIBSE (1992) *The Outdoor Environment*, Lighting Guide LG6.
[3]Lennox Moyer, J. (1992) *The Landscape Lighting Book*, John Wiley & Sons; ISBN 0 4715 2726 2.

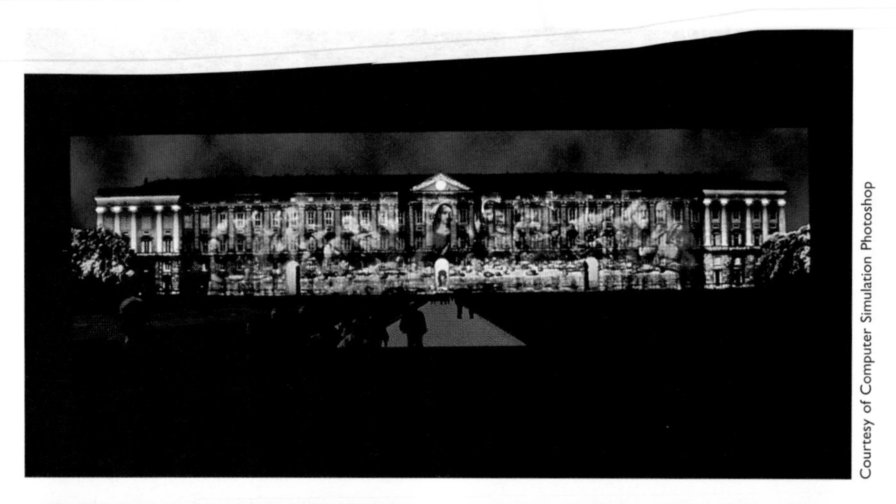

Courtesy of Computer Simulation Photoshop

'Son et lumière' at the Palazzo Reale Caserta Italy. Images of Leonardo's 'last Supper' projected on to the face of a building (Lighting Designer Jonathan Speirs)

Courtesy of International Lighting Review

Lighting up London. The floodlighting of areas of docklands as a 'light show' (Lighting designer Jean Paul Jarré)

Tate Modern, London. Special dynamic lighting for the opening of the building 2000. Opening night (Lighting designer Imagination)

Courtesy of Martin Professional

Alton Towers. Projected lighting effects (Lighting designer Martin Professional)

Courtesy of Sutton Vane Associates

St Paul's Cathedral, London, red floodlighting to recognize 'World Aids Day' on December 1st each year (Lighting Designer Sutton Vane Associates)

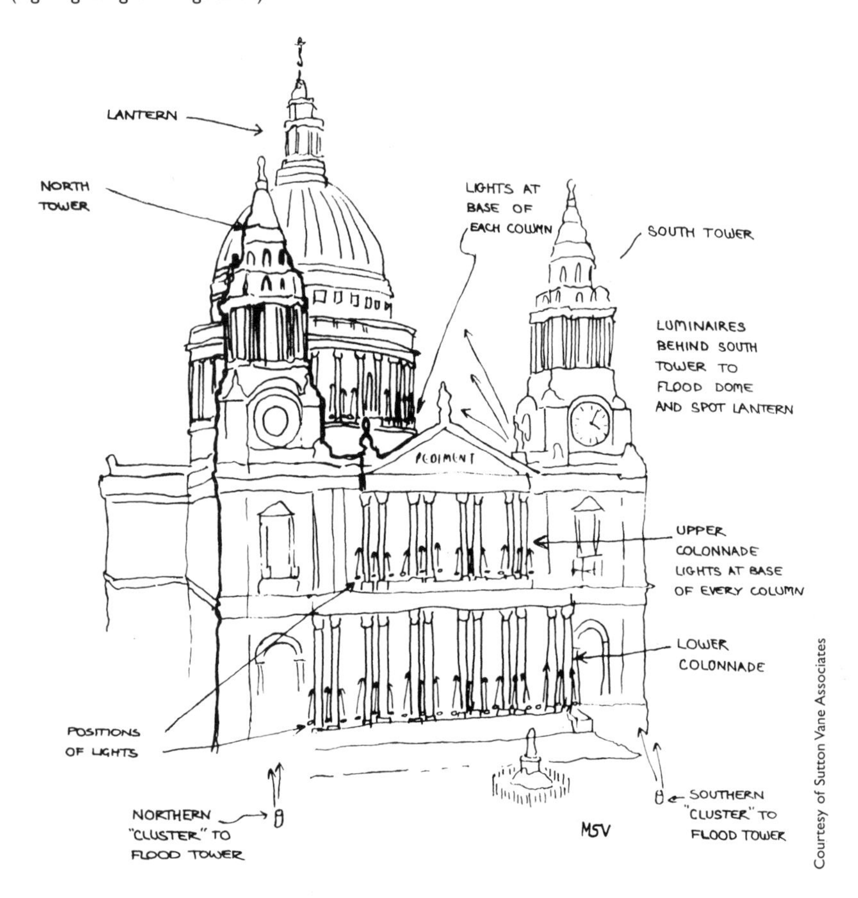

Courtesy of Sutton Vane Associates

Diagram of St Paul's Cathedral (Lighting Designer Sutton Vane Associates)

Courtesy of Allan Toft for Martin Professional

Burj al Arab Hotel, Dubai. Coloured projectors used to alter the colour of the façade (Lighting designer: interior Maurice Brill; exterior Johathan Speirs)

Courtesy of Herbie Knott

The Eden Project, Cornwall. Refer to 'Buildings', page 40

others. It would have to include the natural ephemera of a golden sunset or the first fall of snow.

All that needs to be done in this short section is to illustrate a few examples where a designer has produced an ephemera for some specific reason, using the exciting palette of modern lighting technology to achieve his effect. As the technology develops so will the infinite possibilities to create transitory magic.

Part 3

5 Tactics

Lighting decisions ... illumination level ... the light source ... the fitting ... location ... lighting columns ... wall-mounted lanterns ... lighting bollards ... floodlights ... costs/energy ... interior lighting ... security/emergency ... colour/advertising

Having established the need for a scheme of nighttime lighting, and the most appropriate appearance and views, the tactics to be adopted must be agreed.

Tactics can be divided into four main design decisions as follows:

1. The illumination level required
2. The most appropriate light source: the lamp
3. The best lighting equipment: the fitting
4. The location of the equipment: access.

There are other aspects which may affect these major decisions: such as the ambience of the location, energy considerations, the length of time that the installation will be used and the resources available; all of which may affect the four major areas of choice, and will need to be considered.

Each of the four major areas of decision is discussed below.

ILLUMINATION LEVEL

The level of lighting must be specified, the intensity of the light provided, and the reflectance from the surfaces of the building materials or other exterior surfaces, which will influence the final appearance of the scheme.

It should never be forgotten that 'a little light goes a long way at night', and there is a tendency to overestimate the amount of light required. A salutory example of this was during 1999, when the Church Floodlighting Trust funded by the Millennium Commission, was entrusted with the task of floodlighting some 400 churches throughout the British Isles for the millennium.

Floodlighting schemes were put forward by both consultants and lighting companies, and it was found that in most cases the lighting levels which would have resulted from the proposals were too high, and these were reduced to achieve satisfactory results.

The CIBSE Lighting Guide LG6, *The Outdoor Environment*,[1] gives detailed advice on light levels for different types of building, with variations for areas of low, medium and high brightness, taking account of the reflectance of the building materials. This is a useful starting point (see Table 5.1).

It can be seen from Table 5.1 that there are very wide differences in the recommendations, e.g.:

15 lux for 'clean white brickwork' in a low ambient environment.
450 lux for 'red brickwork' in dirty conditions in high ambience.

Table 5.1 stresses that this is for preliminary design – while experience suggests that the recommendations err on the high side. Because of this, the need for site trials is of the greatest importance.

Table 5.1 Building floodlighting: suggested illuminance for use in preliminary design

Approximate reflectance	*Typical materials**	*Surface condition*	*Illuminance (lux)/district brightness†*		
			Low	*Medium*	*High*
0.8	White brick	Clean	15	25	40
		Fairly clean	20	35	60
		Fairly dirty	45	75	120
0.6	Portland stone	Clean	20	35	60
		Fairly clean	35	55	90
		Fairly dirty	65	110	180
0.4	Middle stone, medium concrete	Clean	30	50	80
		Fairly clean	45	75	120
		Fairly dirty	90	150	240
0.3	Dark stone	Clean	40	60	100
		Fairly clean	55	90	150
		Fairly dirty	110	180	300
0.2	Granite, red brick	Clean	55	90	150
		Fairly clean	80	140	230
		Fairly dirty	160	280	450

CIBSE Lighting Guide LG6 (1992).
*Based on reflectance for a white light. Values may differ if low pressure sodium or high pressure mercury lamps, or lamps emitting light of a predominant colour are used; illuminance should then be decided by site trials.
†Typical districts are: low brightness–rural; medium brightness–suburban; high brightness–town and city centres.

Likewise, with the exterior ground surfaces, the levels of light need to be related to the particular circumstances of the surrounding environment, since at low adaptation levels a person can see sufficiently well in comparatively low levels of light. The danger lies where there is too great a contrast between the levels of light in adjacent areas, the darker areas being obscured where the eye adjusts to the higher levels nearby.

[1]CIBSE (1992) *The Outdoor Environment*, Lighting Guide LG6.

Table 5.2 Typical recommended exterior nighttime illuminances

Purpose	*Lux*	*On which surface?*
Driveways to buildings, secondary pathways	5	Horizontal with at least 50% this value on vertical
Main paths, outdoor car parks	10	Horizontal with at least 50% this value on vertical
Security areas around buildings, main shopping streets	20	Vertical at 1.5 m above-ground
Steps, footbridges and similar hazardous zones, entrance areas to buildings, recreational football pitches	50	Horizontal (vertical surfaces of steps should be differentiated)
Covered pedestrian areas, arcades	75	Vertical at 1.5 m above-ground
Illuminated signs in low brightness districts	100	Vertical
Bus stops, coach loading areas, recreational tennis courts	150	Horizontal
Illuminated signs in high-brightness districts	500	Vertical

The Design of Lighting, Peter Tregenza and David Loe, E & FN Spon, 1998.

Illuminance recommendations for exterior ground areas are suggested in Table 5.2. It should be noted that in high risk areas the levels may need to be raised to as much as 400 Lux, but this will be in very exceptional circumstances where the environmental brightness in city centres is high. Where CCTV cameras are in use, illuminances must relate to those demanded by the camera.

LAMPS

The choice of a suitable lamp must be made since a wide variety of lamps may have some relevance to exterior lighting; but it is the combination of the lamp and the light fitting which must be decided.

This means, in the first instance, the choice of light source – the lamp – to determine the colour relationship with the materials of construction of the building; and in the second, the lighting techniques most appropriate to the particular problems presented by the structure.

There are many useful references available from both the light fittings manufacturers, and the lighting institutions to assist the architect when making his choice of lamp to fit the given circumstances. The following aspects are relevant to his choice. These are not given in any order of importance, since what may be of importance for one project may be less important for another.[2]

Lamps guides:
- Manufacturer's literature
- CIBSE, LG6, Exterior lighting guide
- CIBSE, LG4, Sports
- LIF Lamp Guide
- Lighting the environment.

---

[2]CIBSE/ILE (1995). *A Guide to Good Urban Lighting.*

Wattage and size

Lamps vary considerably in available wattages, there being a relationship with the physical size of the lamp. Some lamp types, particularly those of larger wattage will generally require larger fittings, an aspect which may have importance in terms of equipment concealment. The burning position of some lamps is critical, and manufacturer's advice must be sought.

Distribution

Lamp distribution can be divided into two main types: the sealed beam reflector having its own unique distribution, and the lamp associated with an external reflector, which determines how the light from the lamp is distributed.

Efficacy (efficiency)

Closely related to the question of energy, efficacy indicates how efficiently the lamp converts electrical energy into light, the most efficient (for example, in the range of discharge lamps, providing over 100 Lumens per watt (Lm/W)). But, since discharge lamps require control gear to operate, the energy dissipated by this gear must also be taken into account, since to some extent this reduces the overall efficiency of the lamp circuit.

In today's low energy culture the efficacy should be carefully considered when choosing the lamp. Where the hours of use are likely to be low, it may well be that the simpler, less efficient lamps such as tungsten halogen, may be appropriate; but where there are likely to be long hours of use, a calculation must be made between the higher initial cost of energy efficient lamps, and the energy savings involved taking into account their longer life.

Rated life

This is the lamp's life, or how long the lamp will run in controlled conditions, or how much useful remaining light output it has before it fails. This is a figure provided by the manufacturer and needs to be related to both energy and overall cost in use.

An obvious comparison is that of the original tungsten light bulb, with its rated life of 1000 hours, giving an efficacy of around 12 Lm/W when compared with a modern fluorescent lamp with a rated life as high as 16 000 hours and an efficacy of around 80 Lm/W. The cost difference is significant. Conversely, high intensity discharge (HID) lamps generate high efficacy at around 100 Lm/W.

Hours of use

Where the hours of use for a proposed installation (as, for example, in the interior of a church) are few, it may well be that a scheme using relatively expensive energy saving lamps could prove the wrong option. The likely hours of use is therefore a further consideration.

Colour appearance (lamp)

This refers to the colour temperature of the lamp, or light source, which is a measure of its surface appearance, either warm or cool.

Colour temperature is measured in degrees Kelvin, e.g.:

Tungsten filament	2800 degrees Kelvin	Warm
Daylight overcast	6500 degrees Kelvin	Cool

The CIE Colour Rendering Group is a more useful classification of the lamp output which will determine the colour appearance of the object or surface lit by a particular light source. This varies from 1A (excellent) to 4 (poor). It is important that where accurate colour rendition of objects is desirable, a lamp providing a figure of A or 1B is achieved (see Table 5.3).

A different measure of colour appearance is the CRI or Colour Rendering Index on a scale from 0 to 100. Again this refers to the appearance of objects or surfaces lit by the lamp (see Table 5.3).

Examples of this are as follows:

100 excellent, e.g. natural daylight, tungsten filament and some metal halide lamps
85 Very good, e.g. triphosphor fluorescent lamps
60 Fair, e.g. high pressure sodium
50 Fair, e.g. halophosphate fluorescent lamps
20 Poor, e.g. low pressure sodium lamps

Cost

The price of the lamp, if taken in isolation, is only one cost factor, and perhaps the least useful measure, since lamps of the highest initial cost will often be those that provide the best lifetime economics. It is essential that a cost analysis which shows the 'costs in use' is carried out, to take into account the hours of use of the system, the cost of energy, the initial design costs, and the cost of the lighting equipment and its installation; together with the further cost of lamp replacement and maintenance.

Controls

There are basically two types of control:

1. Control gear, for individual lamp operation.
2. System controls, for the whole lighting installation

Control gear

All discharge lamps operate by using some form of control gear. In the case of early fluorescent lamps this would have been comparatively simple and cheap, but modern fluorescents are moving towards the use of hi-tech electronic ballasts as are the more powerful discharge sources. The control gear required to operate each lamp is complex and expensive, and the cost a serious financial consideration.

A further aspect may well be the start-up time required if for some reason the lamp power is switched off, and it is required to energize the lamp again – in some cases this may take up to three or four minutes. The advantage of filament lighting is that it requires no control gear so that

Table 5.3 This table illustrates the different aspects of the main types of lamp, providing comparisons to assist the architect in making his choice. The different factors identified are those of efficacy, lamp life and colour, but other factors that must also be considered are those of cost and control

Lamp	*Type*	*Lamp efficacy (Lm/W)**	*Circuit efficacy (Lm/W)†*	*Rated average life (hr)‡*	*Wattages (W)*	*Colour temp (K)§*	*CIE group¶*	*CRI***
Incandescent	Tungsten	7 to 14	7 to 14	1000	15 to 500	2700	1A	99
	Filament							
	HV Tung. Halogen	16 to 22	16 to 22	2000	25 to 2000	2800 to 3100	1A	99
	LV Tung. Halogen	12 to 26	10 to 25	2000 to 5000	5 to 150	2800 to 3100	1A	99
Discharge	Low pressure sodium (SOX)	100 to 200	85 to 166	16 000	18 to 180	N/A	N/A	N/A
Flourescent tubes	Cold cathode	70	60	35 to 50 000	23 to 40 W/m	2800 to 5000 2	1A 85 to 90	55 to 65
	Halophosphate (T8 & T12)	32 to 86	13 to 77	10 000	15 to 125	3000 to 6500	2 to 3	c. 50
	Triphosphor (T5 & T8)	75 to 104	CCG: 48 to 82 ECG: 71 to 104	10 000 20 000	10 to 70	2700 to 6500	1A & 1B	85 to 98
Compact fluorescent	Triphosphor	40 to 87	CCG: 25 to 63 ECG: 33 to 74	8000 10 000	5 to 57	2700 to 5400	1B	85
Induction (fluorescent)	Triphosphor	65 to 86	60 to 80	60 000 (service life)	55 to 150	2700 to 4000	1B	85
High pressure discharge	High pressure sodium (SON)	75 to 150	60 to 140	28 000	50 to 1000	1900 to 2300	2 & 4	23 to 60
High pressure discharge (not recommended for new installations)	High pressure mercury (MBF)	32 to 60	25 to 56	24 000	50 to 1000	3300 to 4200	2 & 3	31 to 57
High pressure discharge	Metal halide							
	(quartz)	60 to 120	44 to 115	3000 to 15 000	35 to 2000	3000 to 6000	1A to 2	60 to 93
	(ceramic)	87 to 95	71 to 82	9000 to 12 000	35 to 150	3000 to 4000	1A to 2	80 to 92

Lighting Modern Buildings, Osram Lighting, Updated to June 2001.

*Lamp efficacy indicates how well the lamp converts electrical power into light. It is always expressed in Lumens per Watt (Lm/W)

†Circuit efficacy takes into account the power losses of any control gear used to operate the lamps and is also expressed in Lm/W

‡Rated average life is the time to which 50% of the lamps in an installation can be expected to have failed. For discharge and fluorescent lamps, the light output declines with burning hours and is generally more economic to group replace lamps before significant numbers of failures occur

§Colour temperature is a measure of how 'warm' or 'cold' the light source appears. It is always expressed in Kelvin (K), e.g. warm white 3000 K, cool white 4000 K

¶CIE colour rendering groups: A (excellent); 1B (very good); 2 (fairly good); 3 (satisfactory); 4 (poor)

**CIE colour rendering index: scale 0 to 100 where: 100 (excellent, e.g. natural daylight); 85 (very good, e.g. triphosphor tubes); 50 (fairly good, e.g. halophosphate tubes); 20 (poor, e.g. high pressure sodium lamps)

In the case of reflector lamps, where the light output is directional, luminous performance is generally expressed as *Intensity* – the unit of which is the *Candela* (Cd) (1 Candela is an intensity produced by 1 Lumen emitting through unit solid angle, i.e. Steradian)

the circuit wiring can be simplified, and simple dimmers used for raising and lowering the intensity of light. Illumination is immediate when the lamp is switched on.

System controls

Not to be confused with the control gear, system controls may be provided for raising and lowering light levels, relating light levels to external natural conditions, switching systems on or off at desired times, or modifying colour and the like. The development of control systems in recent years has been at the heart of energy saving, and for achieving the sort of artistic variation for which in the past only a theatre lighting system would have been capable.

Architects are well aware of control of lighting for the interior of buildings related to the Building Energy Management System (BEMS) and exterior lighting systems work in a similar manner; either by means of time control, photocells, dimming, scene sets, or sophisticated colour change methods, whilst even the distribution of the light from groups of fittings can be varied. System controls may be set up to change from time of day, the seasons of the year, or the nature of the weather. Specialist advice is essential, and the installations do not come cheap.

LIGHTING EQUIPMENT

Having established an order of priority for the choice of lamp, it will then be necessary to source appropriate fittings. Exterior light fittings (luminaires) are available from certain manufacturers, but some will be more appropriate than others, and it is important to establish a general specification to avoid problems which can arise.

The architect will no doubt require assistance with the design of an exterior floodlighting scheme, and the International Association of Lighting Designers (IALD, Merchandise Mart Suite 11–114A, 200 World Trade Center, Chicago, Illinois 60654, USA) provides the only worldwide list of 'independent' lighting design organizations.

Fittings manufacturers on the other hand provide a useful design service to illustrate the light distribution of their products related to a building by computer plots. However, this approach does not obviate the need for a lighting trial to illustrate the visual aspects of the scheme to both the design team and to the client. Lighting trials can do so much more than computer plots – which tend to be limited to information of light levels – for while limited by economics to specific areas of a building, they can be set up to show the overall composition, the modelling, and colour. These can then be judged from close-up or distant views.

Optics

The light distribution from the fitting is determined by the combination of lamp with its reflector, except in the case of sealed beam lamps. Fittings will be designed to provide a variety of light distributions, such as symmetrical, or asymmetrical, narrow or wide beam.

In most instances, floodlights will require accurate mounting to ensure that they deliver the light to the design distribution, not only on initial installation, but also during relamping or when the fittings are cleaned. Special aiming arrangements are provided on many fittings to ensure that this can be carried out. This is of particular importance where fittings are aimed upwards at buildings to ensure a total cut off of light at the roof line, to prevent spill light shining up to the sky as light pollution.

Safety and quality

The first and most obvious consideration is that any fitting must comply with the appropriate British Standards. Technical standards are now international, so that British Standards are identical to the European equivalents. Many light fittings carry the European approval mark ENEC, which means that they have been tested and approved for safety and manufacturing quality and provide quality assurance. While this will normally be available from UK manufacturers and reputable European manufacturers, it is not always the case with imported fittings, and although many products carry the CE Mark to comply with the law, this is insufficient proof of quality or safety.

Classification (IP)

Fittings are classified as to their ability to withstand external conditions. This is known as the 'Ingress protection' covering the ingress of water or dust, and is classified by an IP rating. This consists of two numbers, the first concerned with dust (solid particles) and second with moisture (liquids), e.g. a suitable IP rating for an exterior fitting might be IP65, being dust tight, and giving protection against water jets (Table 5.4). Manufacturers will state the IP rating for their external fittings.

LOCATION

Having made the necessary aesthetic judgement as to how the building should be revealed at night – the right emphasis, the modelling, and the shadows – the location of the equipment to achieve this effect is a vital factor in the designer's choice.

It is arguable that this decision should have been considered before the others – since the choice of where it is practical for the floodlighting equipment to be placed to provide the desired light to the surfaces and edges of the building is critical to its success. There may even be circumstances in which the only feasible locations from a lighting point of view are unsuitable for some other reason, requiring a change of strategy. Much floodlighting equipment is large and cumbersome although, in order to meet the need for delivering sufficient light of the right intensity to the faces of the building, and the question of how this should or could be concealed, this aspect is often not fully assessed.

There may well be situations where it is just not possible to place a fitting in a desired location; because of physical constraints, unacceptable glare, or where the appearance of the equipment would be obtrusive by day. Even allowing for the fact that this may diminish the visual impact of the lit building at night, the deleterious impact of views of the fitting may prove far worse, and perhaps the concept of 'when in doubt leave it

Table 5.4 The degrees of protection against ingress of solid bodies (first characteristic numeral) and moisture (second characteristic numeral) in the Ingress Protection (IP) system of luminaire classification

First characteristic numeral	*Degree of protection and short description*	*Details of solid objects which will be 'excluded' from luminaire*
0	Non-protected	No special protection
	Protected against solid objects greater than 50 mm	A large surface of the body, such as a hand (but no protection against deliberate access); solid objects exceeding 50 mm in diameter
2	Protected against solid objects greater than 12 mm	Fingers or similar objects not exceeding 80 mm in length; solid objects exceeding 12 mm in diameter
3	Protected against solid objects greater than 2.5 mm	Tools, wires, etc. of diameter or thickness greater than 2.5 mm; solid objects exceeding 2.5 mm in diameter
4	Protected against solid objects greater than 1.0 mm	Wires or strips of thickness greater than 1.0 mm; solid objects exceeding 1.0 mm in diameter
5	Dust-protected	Ingress of dust is not totally prevented but dust does not enter in sufficient quantity to interfere with the satisfactory operation of the equipment
6	Dust-tight	No ingress of dust

Second characteristic numeral	*Degree of protection and short description*	*Details of the type of protection from moisture provided by the luminaire*
0	Non-protected	No special protection
1	Protected against dripping water	Dripping water (vertically falling drops) shall have no harmful effect
	Protected against dripping water when tilted up to 15 degrees	Vertically dripping water shall have no harmful effect when the luminaire is tilted at any angle up to 15 degrees from its normal position
3	Protected against spraying water	Water falling as spray at an angle up to 60 degrees from the vertical shall have no harmful effect
4	Protected against splashing water	Water splashed against the enclosure from any direction shall have no harmful effect
5	Protected against water jets	Water projected by a nozzle against the enclosure from any direction shall have no harmful effect
6	Protected against heavy seas	Water from heavy seas or water projected in powerful jets shall not enter the luminaire in harmful quantities
7	Protected against the effects of immersion	Ingress of water in a harmful quantity shall not be possible when the luminaire is immersed in water under defined conditions of pressure and time
8	Protected against submersion	The equipment is suitable for continuous submersion in water under conditions which shall be specified by the manufacturer

CIBSE, *Code for Interior Lighting*, 1994; Classification for 'Ingress Protection'

out' should be considered. Historic buildings present a particular challenge in this respect.

Decisions on the location of equipment must also take into account the question of ease of access, first for maintenance and second to avoid the possibility of vandalism.

Some illustrations of location, both good and bad are given here.

While it is desirable where possible, for the lighting equipment to be concealed, this is not always possible and the daytime appearance of light fittings for exterior use at night is an important element of choice. It is sometimes helpful for the fittings to be finished in a colour appropriate to their surroundings.

There appear to be two schools of thought: on the one hand, those that believe that anything which is functional (and exterior lighting must be functional if it is to be anything) should eschew decoration of any sort; and on the other, that appropriate decoration in keeping with its surroundings is not only permissible but mandatory.

While there are still old buildings, there will still be a call for heritage fittings; but functional fittings, if they are true to themselves, can also be

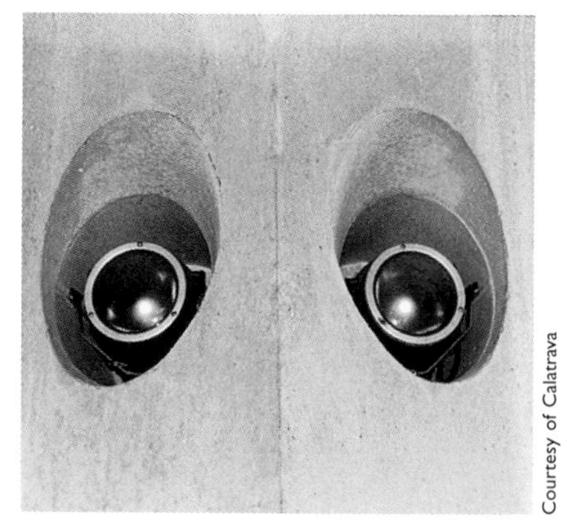
Courtesy of Calatrava

Comparison of the mounting of equipment lighting a railway station in Lyon by Calatrava, where the fittings have been recessed into the structure

DP Archive

Inappropriate and very visible mounting of equipment lighting a major bridge on the Thames

DP Archive

Concealed floodlights lighting the old Abbey at Bisham

Brandston

Statue of Liberty fittings below metal louvre grid

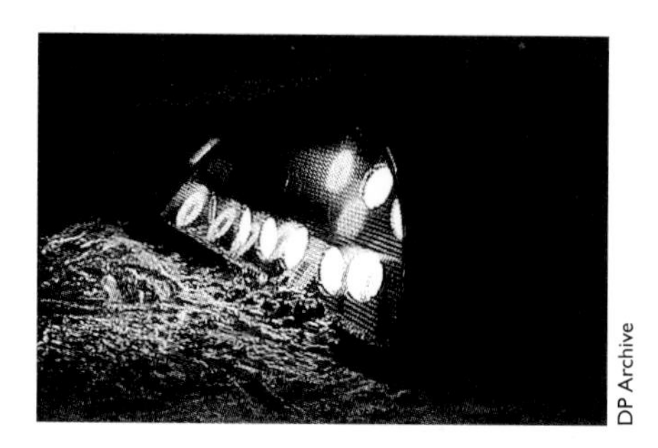
DP Archive

Albert Memorial, concealed fittings

DP Archive

Inappropriate Floodlights at St. Paul's, lighting towards seating by day

DP Archive

Inappropriate Floodlights at St. Paul's, lighting towards seating by night

DP Archive

Well-concealed floodlights at the Philips Lighting Centre in Croydon

appropriate. This is a little like putting a 'barcelona chair' in a 'louis quatorze interior' . . . it can work!

EXTERIOR FITTING TYPES

There are five generic types of exterior lighting equipment:

1. Column mounted fittings: used for streets, car parking, cycle tracks and pedestrian areas
2. Wall mounted lanterns: used for road and pathway lighting
3. Lighting bollards: used for pathway and area lighting
4. Floodlights: used for floodlighting buildings, bridges, statues, etc.
5. Recessed fittings: used for pathways, or where buried in the ground for uplighting structures, trees and other incidents.

Exterior lighting equipment is in a constant state of development, and it is therefore impossible in a book of this sort to attempt to illustrate a current range of equipment which will be appropriate even at the time of publication. What is attempted, is to make some observations which may be of help to those seeking advice about the different types of equipment, illustrated by some examples.

Lighting columns

The importance lies not so much in the column, but the lantern it supports. The height of the column and the distance the columns are placed apart determine the level of lighting required for the different categories of road. Having said this it does not preclude the design of the column itself which must be related to its surroundings. Due to local authority economics, the presumption in the UK has been to maximize the light output in a sideways direction, in order to stretch the light as far as possible and thus reduce the number of lanterns to a minimum. While this is laudible in itself it tends to lead to glaring road conditions. Conversely, in other European countries, there is a greater use of cut-off lanterns designed to control the light downwards reducing glare, but requiring columns to be set at closer distances. This tends to lead to pools of light more appropriate for the use of pedestrians than for cars.

There is now a move to separate the lighting needed for the car from that required for the pedestrian, with multi-use columns, where the lantern at the top of the column is directed towards the road, with a smaller fitting mounted at a lower level directed towards the pathway. The spacing of the lower pedestrian lamps dictated by the spacing of the higher road lighting tends to provide pools of light, and these are sometimes supplemented by shorter columns between – providing infill light – although these will generally not be necessary. Where a systematic approach to differentiate light colour for roads and people has been planned, the lower lamps may be of a different (warmer) colour to the higher.

In the case of lighting for squares and parking areas, the columns do not need to be so high, and cut off lanterns are appropriate to avoid glare. At one time it was popular in some commercial car parking areas, such as supermarkets, for glass or polycarbonate globes to be placed on short columns, giving all-round light. These were a glaring disaster, and in the main have disappeared.

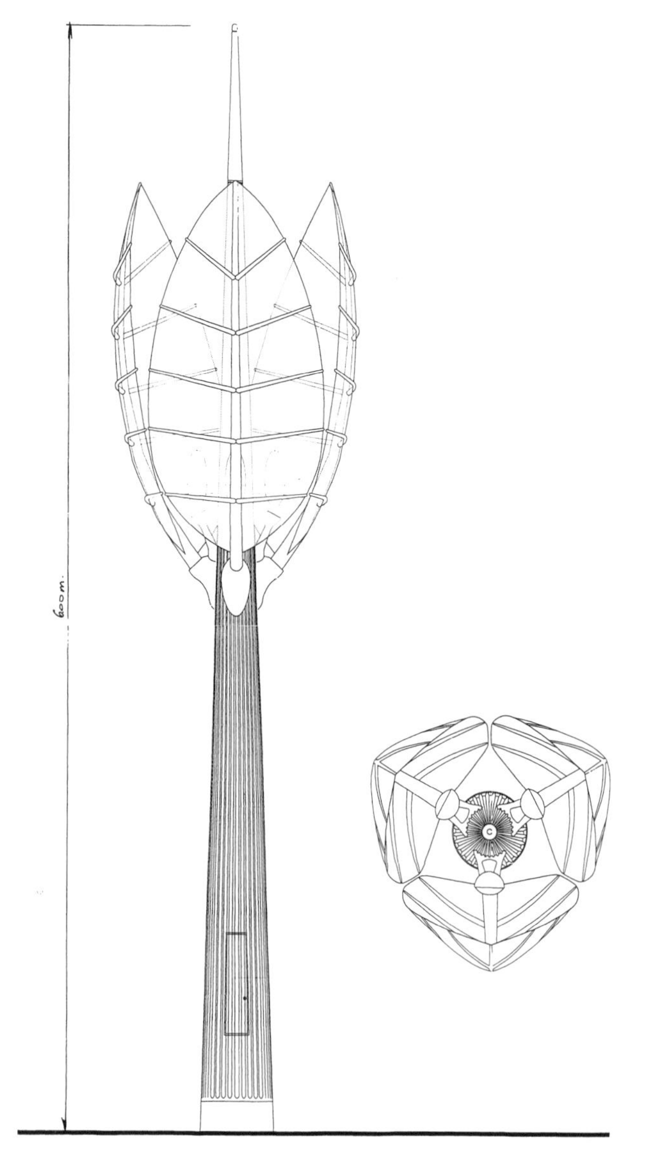

Courtesy of Ville de Lyon

Le Mat Tulipe, Decorative lantern

Courtesy of DW Windsor

The 'blue' lantern, Bluewater

Courtesy of Geo

The 'Park light'

Courtesy of Louis Poulson

Airport 2000

Courtesy of LB Lighting

Heritage fitting

Courtesy of DW Windsor

Stoke Luminaire

There is an opportunity for those wishing to establish a particular identity to create innovative lantern designs such as that for 'Le Mat Tulipe' in Lyon, or the Blue Lantern for the shopping centre at Bluewater. A further development in lanterns is to use the indirect light reflected from a canopy placed above the light source, an example being that of the 'Park Light' from Geo, which, because of its low glare downward light, is most suitable for pedestrian or car parking areas.

Wall-mounted lanterns

Because of the proliferation of street lamps, directional signs and advertising, it was thought that one way of diminishing the clutter was by eliminating one of the culprits: the street light. This was achieved by placing the lanterns on the walls of adjacent buildings on either side of the road. This transferred the problem from the street level to the buildings, where the lanterns tended to create confusing light patterns on the building façades.

There is certainly a place for wall-mounted lanterns, but these need to be fittings with carefully controlled distribution, eliminating unwelcome spill light, and can be successful in low ambience situations providing gentle pools of light suitable for pedestrians.

The wall brackets at Lambeth Palace in London are an example of successful wall lights designed to fit a heritage situation. A more usual type is the small fitting mounted in a wall recess giving a degree of light to a pathway; fittings varying from the simple bricklight to more elaborate distributions. There is always the danger that the machine-made fitting will tend to be incompatible with the man-made brick or concrete structure, and this can be overcome by a loose-fit approach, or a wide flange to the fitting covering the edges of the structure. Surface-mounted wall lights are still widely used, but this method should be used with care, as it can be obtrusive to the architecture, and provide conflicting light distribution.

Lighting bollards

It is perhaps surprising that the simple task performed by a lighting bollard – that of lighting a small surrounding area or pathway – should have produced such a multitude of design solutions.

The task of providing all-round light is typified by the solution used for Whitehall, Leeds. While a multi-purpose solution is illustrated by the street furniture for Trafalgar Square, where in addition to acting as a lit bollard, the fitting houses floodlights which light up to the base of Nelson's column. The technology for this is reminiscent of the lighting columns in the Place Royale in Brussels designed many years ago, where the tops of traditional street lighting lanterns were modified to incorporate upward lighting floods to the face of the Town Hall; a solution often copied by others.

The magic of gas-light dies hard, as shown by the flambeau outside the Lanesborough Hotel in London. Yet another twist to the lit bollard story is given by the lighting standards designed as lit sculpture; which by variations in height can be formed into groups, still performing the job of lighting bollards (see Hilton Hotel, Tel Aviv., Lit Environment p. 152).

Confusing patterns on buildings

Wall bracket at Lambeth Palace

Bingo Club, Brighton

Recessed wall mounted fitting

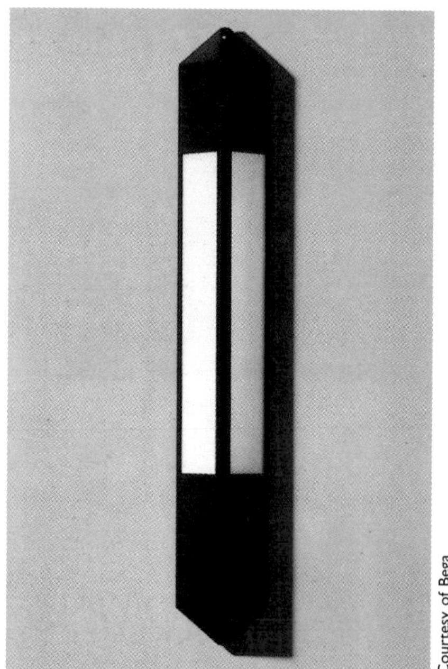

Fluorescent wall light

Courtesy of Urbis

Whitehall, Leeds

Courtesy of DZ Licht

Group of sculptural bollards

Courtesy of Sugg

Trafalgar Square multi-purpose bollards

Courtesy of Sugg

Flambeau fittings, Lanesborough Hotel, Hyde Park

Courtesy of Thorn

Bollard

Floodlights

In a little pamphlet *Floodlighting Buildings*, produced by the RIBA in 1983 for its 150th Anniversary, flood lights were divided into three categories:

A. Floodlights giving a symmetrical beam
B. Floodlights giving a fan-shaped beam from a linear source
C. Floodlights similar to B, but giving an asymmetrical beam, distributed vertically.

Despite all the advances in the design of floodlights, and the fact that these categories are no longer used, the situation remains much the same. Floodlights have a distribution which is related to the shape of the reflector, and to the relative position of the lamp. A round reflector produces a conical beam. This symmetrical type is useful for picking out detail either close to or at a distance, and it should be remembered that if the beam is at an oblique angle the result will light an elliptical area.

On the other hand, an asymmetrical reflector is useful for lighting areas, whilst a double asymmetrical reflector provides an offset to the vertical output.

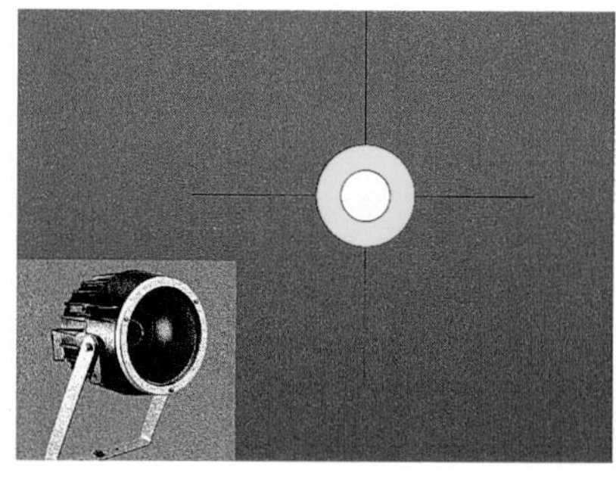

Sketch of symmetrical distribution of long throw beams (Type A)

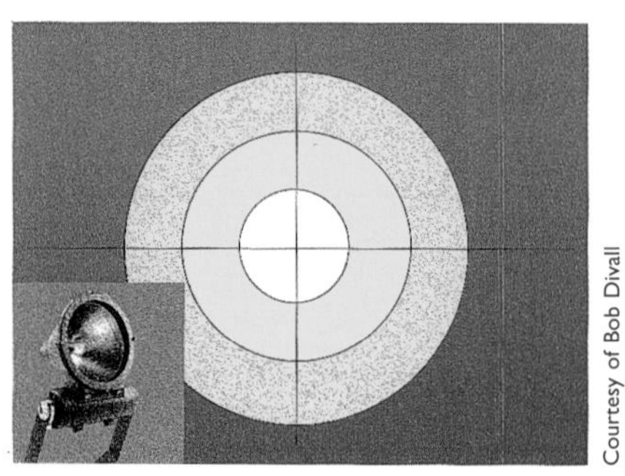

Sketch of symmetrical distribution of short throw beams (Type A)

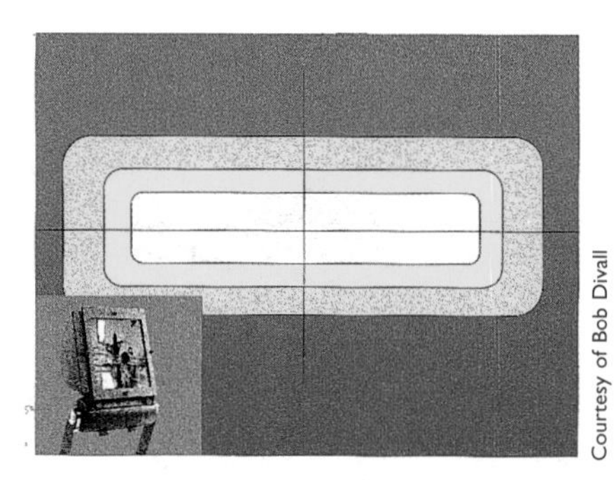

Asymmetrical distribution for lighting areas (Type B)

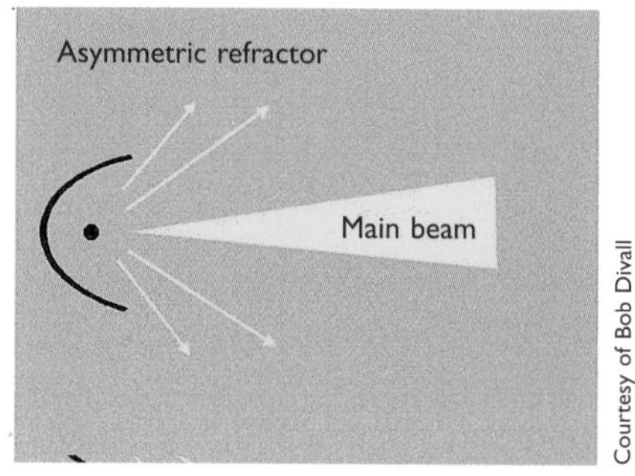

Asymmetrical reflector for lighting areas (Type B)

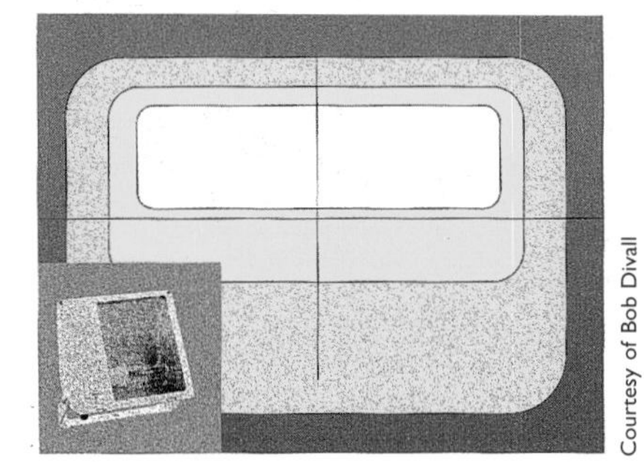

Double asymmetrical distribution for vertical 'offset' (Type C)

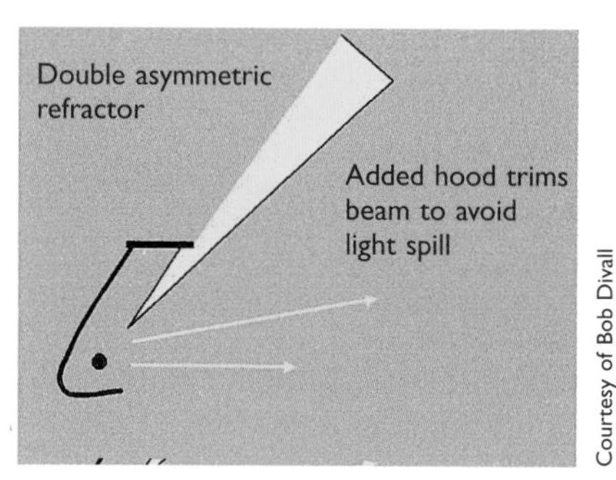

Double asymmetrical reflector, for vertical 'off-set' (Type C)

It is more often the case that off-the-peg floodlights will produce the distribution required for a specific setting. With modifications in the form of refractor glasses and cut-off shields or louvres applied where necessary to provide precise beam distribution and prevent light spill. These types of floods are the all-important workhorses of exterior lighting, and they are in a constant state of improvement and development.

FLOOD LIGHTING TECHNIQUES

Diagrams of location

1. FRONTAL LIGHT

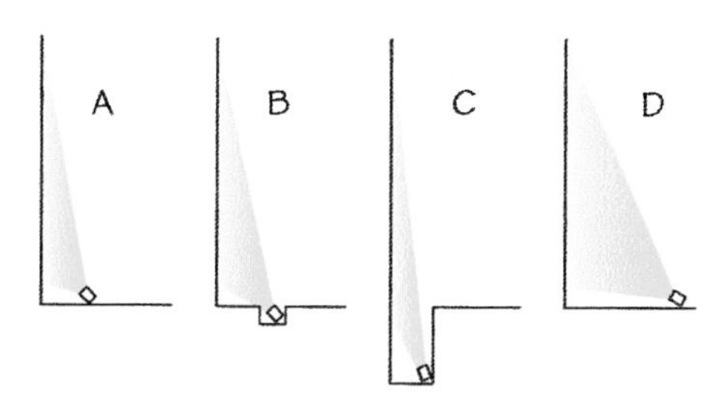

2. LIGHT FROM HIGH LEVEL AT A DISTANCE

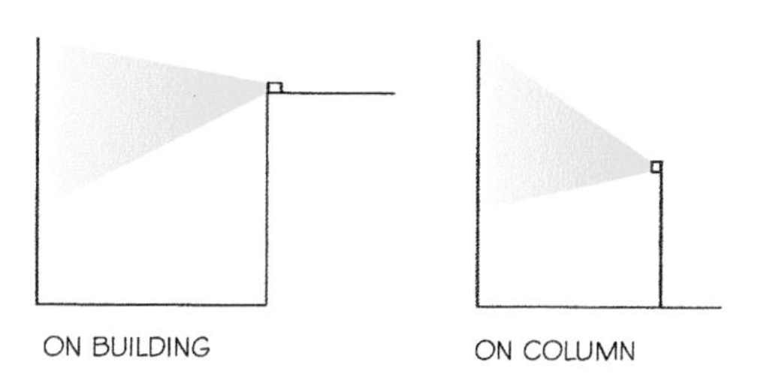

3. LIGHT FROM THE BUILDING

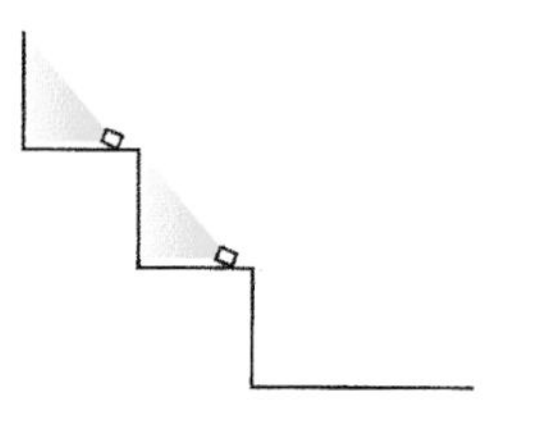

4. INTERIOR LIGHTING

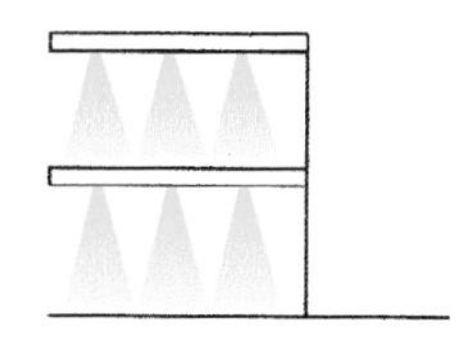

Courtesy of Mark Sutton Vane

Tsola fitting (solar powered)

DP Archive

Buried flood St. Albans Church

The way in which floodlights are located, in relation to buildings, are illustrated above: (1)–(3) shows external floodlighting, while (4) shows a building lit from its own internal lighting.

Recessed fittings

The concept of recessing floodlights into the ground is comparatively new, but is now recognized as a mainstream solution, with fittings used for the upward lighting of trees, sculptures and the sides of buildings. In the past there have been problems of overheating and the ingress of water, but these have been mainly overcome, and there is little doubt that the high quality fittings now offered are satisfactory.

DP Archive

Treelighting, Port St Louis, France

COSTS/ENERGY

The subject of costs has already been introduced, but since this is an important part of the decision-making process, it demands further exploration.

With new building work the question of exterior lighting needs discussion at the design stage, depending upon the type of building, as to whether the interior lighting is such as to make exterior floodlighting superfluous, or whether some form of partial exterior lighting would suffice. It is to be hoped that the architect will have considered the second aspect or the appearance of the building at night, at the design stage, and will be in a position to discuss the philosophy behind his or her design.

It is worth considering the exterior electrical installation which may be required, since if put in at the construction stage, it will be less costly than if left until the building is complete.

The situation is different with an existing building, which may be of historic interest, where a decision has been made – perhaps as a part of the visual masterplan for an area – to light it at night. Here there may be no choice but to consider exterior floodlighting; although, as the lit windows at the Valencia bull ring indicates, there can be alternatives (see Chapter 2 and p. 27).

Having made the decision to light the building in one way or another, the location will be known and a decision has to be made in reference to the surroundings as to the illuminance which would be appropriate to the site.

Table 5.1 has indicated the range of light levels appropriate to the nature of the surroundings, whether town or countryside, but it should always be remembered that unless the building is competing with the centre of Las Vegas, it will be better to err on the side of less rather than more light.

Exterior lighting has become a great deal more sophisticated now than in the days when it was thought sufficient to throw a great deal of light from a frontal direction on to a façade, ending up with the building looking like a cardboard cut-out – this was disastrous aesthetically, and damaging environmentally with a waste of light and consequently energy.

Wasted energy is associated with light pollution and light trespass, two sides of the same coin. Lighting which spills out beyond the surface it is designed to light causes pollution to the sky, and may add trespass to the windows of neighbouring houses. It also wastes energy and is to be deplored.

Once the parameters are known, the essential cost appraisal should be made before any work is started, so that the building owner is aware of the initial costs, and the operator aware of the estimated running costs. It is assumed that the capital costs of alternative designs will be assessed in relation to their running costs.

Capital costs

- The costs of design
- The initial cost of the lamps, fittings and fixed light controls (louvres, etc.)
- Cost of any necessary electronic control systems
- The cost of the electrical installation
- Cost of any associated building work/anti vandalism
- The interest on capital employed.

Running costs

- Estimate of the annual cost of electricity (taking into account the annual hours of use)
- The annual cost of lamp replacement
- Maintenance cost for cleaning.

It is generally best to keep estimates of the capital and running costs separate, since these may well have to be borne by different entities. Although it would, of course, be possible to work out an annual cost for the two together, bearing in mind that while capital costs remain static until such time as the whole system is updated, the running costs tend to rise with inflation.

It will generally be the case that when the capital costs of the nightlighting system are known, these will form a minor part of the overall costs for a project. It is only when they are initially overlooked and have to be added later that problems arise. Furthermore, the costs of energy will form a small part of the overall energy costs of running the building.

INTERIOR LIGHTING

The interior lighting of an historic building may have contributed little to its night appearance, except perhaps in the case of churches where light seen through stained glass produced its own aesthetic, and light from small windows in houses added a domestic flavour to such buildings when floodlit from outside.

The situation with new buildings is very different. The practical and aesthetic needs of daylighting tend to lead to whole glass façades, or translucent structures, where the interior lighting dictates the nighttime appearance of the building, and where it is the nature of the lighting itself which will play a major part in the creation of the unity of the building. A good example of this is in the appearance of multi-storey office developments. Initially, a developer would have built the shell of the different floors, leaving it to the occupier to fill in the various services he required. This had the effect of producing many different solutions to the lighting at ceiling level, in some cases with lines of fluorescent light parallel to the window wall, and in others at right angles; the types of fitting and the lamp colour at variance.

Forgetting about the problems of light pollution, and wasted energy, the effect was unlikely to produce a unified appearance to the structure at night. The nature of office construction has changed, with a much higher specification for the servicing elements, such as computer floors, service ceilings catering for integration of lighting, air control, acoustics and communication systems. When this philosophy is applied to an office building despite multiple occupancy, it tends to bring with it a rationalization of lighting layouts, which contribute to the unity of the structure.

In the case of a building of mixed form – some areas of total glazing and others of solid construction – a balance needs to be struck between the appearance of the interior seen by means of the interior lighting system, and the areas of unlit wall. There is a danger that where the solid areas of wall are completely unlit, addition of some gentle light to establish the presence of the wall, and its interrelationship with the lit areas may be considered. On the contrary, there may well be some circumstances where, for various reasons, it is the architect's choice to leave some areas in decent obscurity.

Shopping streets are tending to disappear in our new or reconstructed towns, being overtaken by shopping centres with internal shopping malls. The situation in shopping centres is that the areas of public access should be lit to the lowest lighting levels consistent with safety, allowing the interior light from the shops themselves to 'speak' to the shopper, and by its nature, to advertise the range and quality of the goods. In existing shopping streets where cars vie with pedestrians, it has never been a good idea to rely on the interior light from shops to illuminate associated pavements due to the lighting being problematic at best. But where cars are not permitted in pedestrian streets, suitable arrangements may be made with the shop owners to leave lights on until late, and in such circumstances additional light may be unnecessary.

It is clear that interior lighting can form an important part of the nighttime appearance of buildings, and must be a consideration when determining the design of the exterior lighting proposals.

SECURITY/EMERGENCY

While road lighting has implications for security – crime is known to fall in areas where the public lighting of roads and open spaces has been improved – this section is concerned more with security and emergency in relation to properties.

Security lighting

It is undeniable that crime is more likely to be committed at night, and that if a property has a well thought-out scheme of exterior lighting, there will be less chance of criminals being attracted to the site. It is equally true that lighting alone will be insufficient, and that this must be associated with the normal deterrents to crime, in terms of locks, bolts and in some instances security systems. The lighting design should aim to assist in increasing the possibility of an intruder being observed, either by the social supervision of neighbours, proximity floods, or in areas of high risk, such as retail premises, by ensuring that any points of entry are well-illuminated at night. Where buildings have night watchmen or special security staff, there will generally be observation by means of CCTV; in these cases, where observation takes place at a distance, the lighting must ensure that an intruder will be sufficiently well lit, either close to the building or at access points to the perimeter of the site.

Each type of area will need to be surveyed properly to assess the risk involved, and a system developed to suit its particular needs.

Emergency lighting

Linked with the exterior security lighting will be that for emergency. The architect will have arranged for the interior of the building to have a mandatory system of emergency lighting which will be energized in the event of a power failure, the means to achieve this depending upon the nature of the electrical installation, whether by back up batteries or generator, etc. What may not be so clear is that there should be a relationship between the interior emergency system and the exterior lighting. It won't be good enough for staff to negotiate a safe interior escape route from the building, only to find, when reaching the outside, that they are cast into outer darkness. Therefore, it is essential that the assembly and external areas immediately adjacent to the escape routes are lit from the emergency system, and are energized at the same time as that internally.

The design of the exterior emergency system must ensure that on escaping from the interior the light is directed away from the exit towards the area beyond, glare towards those emerging must be avoided at all times.

COLOUR/ADVERTISING

It is generally accepted that the use of coloured light in advertising is quite appropriate in towns within normal planning contraints, provided that it does not conflict with traffic signs, and the massive neon signs in the centre of towns such as Las Vegas add an excitement to the scene.

A modest scheme of advertising a leisure centre in a new town

DP Archive

However, the situation is rather different when dealing with the normal lighting of buildings seen at night. Where floodlighting is considered desirable, it is now possible by modern developments in lighting technology to produce a variety of colour effects onto the façades of buildings. These effects can be of solid colour, colour images, and even dynamic patterns. The palette at the disposal of the lighting designer is huge, and will no doubt continue to expand into the future. Early floodlighting had resulted from the effect of the filament light source, providing what was considered to be a natural appearance on the building it was lighting, albeit a warmer impression on brick or some stonework. This formed its own discipline.

Up to the 1950s and particularly with the introduction of *son et lumière*, colour on buildings had been limited to a relationship between the light sources available and the colours of construction, brick, stone, concrete etc. With the advent of discharge sources, with their obvious economic advantages, many of the original schemes were updated to use the new sources. This had some very strange results, with the natural colour of construction changing in a bizarre manner.

The development of the high-pressure sodium lamp (SON) appeared to be the answer, and many new and old schemes were designed to take advantage of the new economics, coupled with a source giving a warm colour rendering to building materials, thought to be acceptable. It was soon found, however, that the tendency to light all buildings in this manner, whatever the colour of the materials, was unsatisfactory.

The latest in the line of floodlighting sources with better and more consistent colour rendering is the metal halide lamp with CDM technology – having the advantage of excellent economics with cool or warm lamps to fit different circumstances. These are said to provide a natural colour appearance to the building surfaces. This is to be welcomed, but no doubt is not the end of the story in a developing technology.

DP Archive

Lloyds Building floodlit at night/blue (Architect Richard Rogers Partners; Lighting designer Imagination)

So where does this leave the question of colour? It is very much a question of aesthetic judgement; but some really awful examples of the indiscriminate use of colour on buildings exist, where the colour of the building and the light source should give one reason to pause.

A suggested philosophy is as follows:

1. The nature of the building materials should be taken into account, in order for permanent systems of floodlighting to express their appearance as natural.

   An example of this is the Lloyds Building in the City of London, where the designer adopted a system of blue floods to express the stainless steel façades. This might have appeared startling when it was first installed, but has proved to be an acceptable longterm solution. The opposite can be said of the green light on the Hoover Building of more traditional construction, which has found little favour, due both to its colour and unacceptable spill-light to the sky.

   The conclusion is that where a permanent system of floodlighting is installed, this should echo the colour of the natural daylight appearance of the building.
2. Colour on buildings is best left to the ephemera.

   An excellent example of this is the annual practice of red light being applied to St Paul's Cathedral in the City of London on 1st December (World AIDS day) in recognition of the 13 million people who have died from the disease (see Chapter 4).

Courtesy of Imagination

Hoover building floodlit at night/green (Lighting designer Imagination)

The conclusion is that anything goes, provided that it is for a limited period. Colour on buildings joining with fireworks, Christmas lights and *son et lumière*, all add to the sum of human enjoyment.

Part 4

6 Lit environment

INTRODUCTION

The previous chapters have analysed the component parts, which make up the lit environment – the buildings, incidents and the spaces between.

This final chapter is designed to bring these all together, and to show that the successful lit environment is achieved when a combination of all or some of these elements are combined to engender delight – it is this whole which is more important than the sum of its parts.

Due to the fragmentation of the design process in our modern towns there has been little coordination between those who are responsible for lighting the streets, the buildings, and the incidents. We have much to learn from our older cities, which have developed over time, and where there is often a unity between the various elements; as we can see in the Place de Terreaux in Lyon.

A happy marriage can be brought about with an inclusive design team instructed to carry through a visual masterplan in which all the various elements may be coordinated.

Lit environments can be at small or large scale, and it is salutory to experience the small environment as seen through the eye of the artist, for example, in the case of Van Gogh, in his picture of the famous Café le Soir, in the Place du Forum in Arles.

Painted over a century ago, it expresses precisely the intimate environment of the café, the lighting being derived from local sources related to the canopy above the tables, which can be contrasted with a photograph of the café taken from the same viewpoint in 2000. Van Gogh's work might yet again be contrasted with a daylight view of the café, painted in 2000.

Yet another picture shows a covered area in the same square in 2000 with small sources related to the canopy. The environments created indicate the simplicity of approach which has changed little in over a hundred years – it clearly has a popular appeal.

Lit environments will generally be of a larger scale, such as those provided by a town square, a university campus, or a commercial or industrial development, but these too will be composed of the same combination of elements, and the one thing which will distinguish the successful lit environment will be its capacity to delight us.

The opposite to delight is well-illustrated by a square in London (featured in the original Waldram lecture) where, in 1990, anyone would

Courtesy of Rijksmuseum/Otterlo

The Café le Soir, Place du Forum, Arles as painted by Van Gogh, 1888

DP Archive

The Café le Soir, Place du Forum, virtually unchanged since 1888, photographed in 2000

DP Archive

A café in the centre of the Place du Forum, 2000

The daylight environment of the Café le Soir seen in a painting by Elan Kutz, 2000

have felt threatened. So what is it that can on the one hand create delight, and on the other make one look around for the machine guns? It is a question of quality.

Quality is so difficult to define, it has much to do with satisfaction with the function of a space. It must be somewhere one would wish to be; and there must be an established need, although it would not necessarily be one that has been developed over the centuries, as typified by this little market square in Venice. As Gertrude Stein once said 'there must be a "there there"'.

Market Square, Venice

DP Archive

In order to illustrate schemes where it is clear that quality has been achieved; some thirty case studies follow, which it is hoped, will demonstrate the variety and delight which can be achieved, often with very simple means, when architects and other members of the design team work together for its achievement.

City centres

Albert Memorial, London

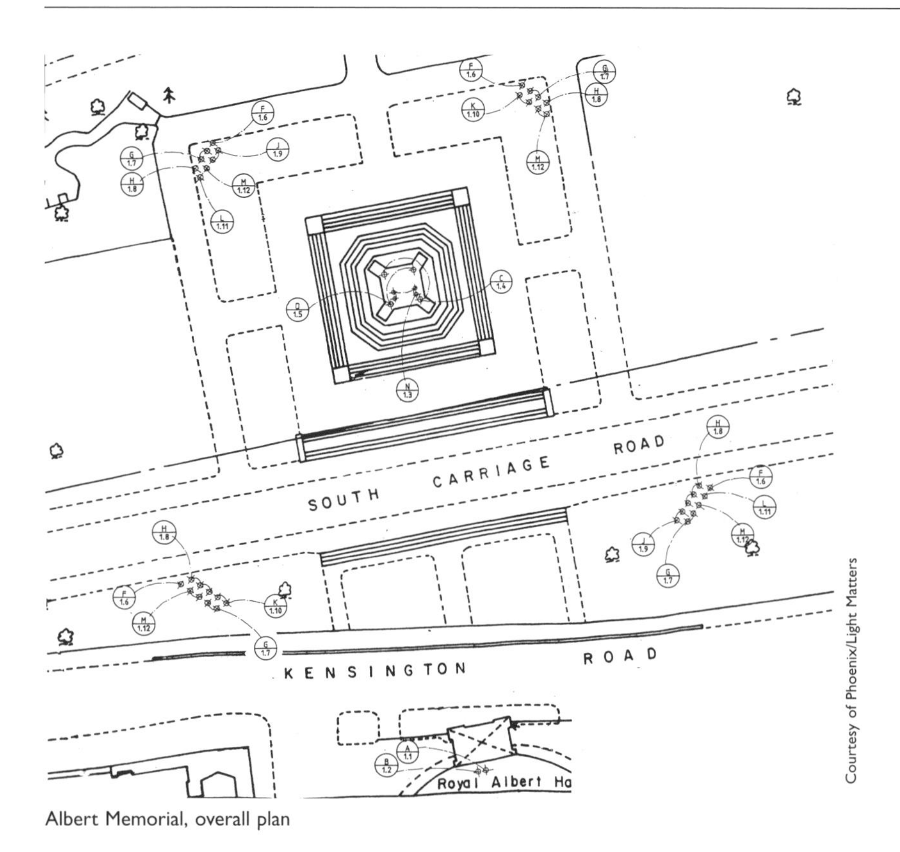

Albert Memorial, overall plan

Courtesy of Phoenix/Light Matters

It might be thought that the lighting of the Albert Memorial, would be better listed under 'sculpture' in the Incidents chapter of Part 2. However, since the overall project includes a wide area of Kensington Gardens in London, set back from the road within the context of the surrounding mature trees, it is more appropriate to include it here.

For many years the Albert Memorial was derided as a symbol of Victorian values and decoration, and the statue had been left in a state of disrepair; it has only been in recent years that the value of such incidents in London has been recognized, and the work carried out to bring it back to its former glory. The lighting scheme was a part of an £11 million restoration of the Memorial carried out by English Heritage in 1998 and it has resulted in the gleaming figure of Prince Albert, set within his memorial architecture 'being one of London's most dramatic new nighttime attractions' (a comment made during the 2000 Civic Trust Award ceremony). Four new corner landscape features were added, which provided lighting pits for lamp concealment for the lower levels whilst the statue of the Prince was lit by two narrow beam search lights mounted on the roof of the Albert Hall. Uplighting was mounted within the historic structure.

A crucial consideration for the designers was the unilateral nature of the lighting

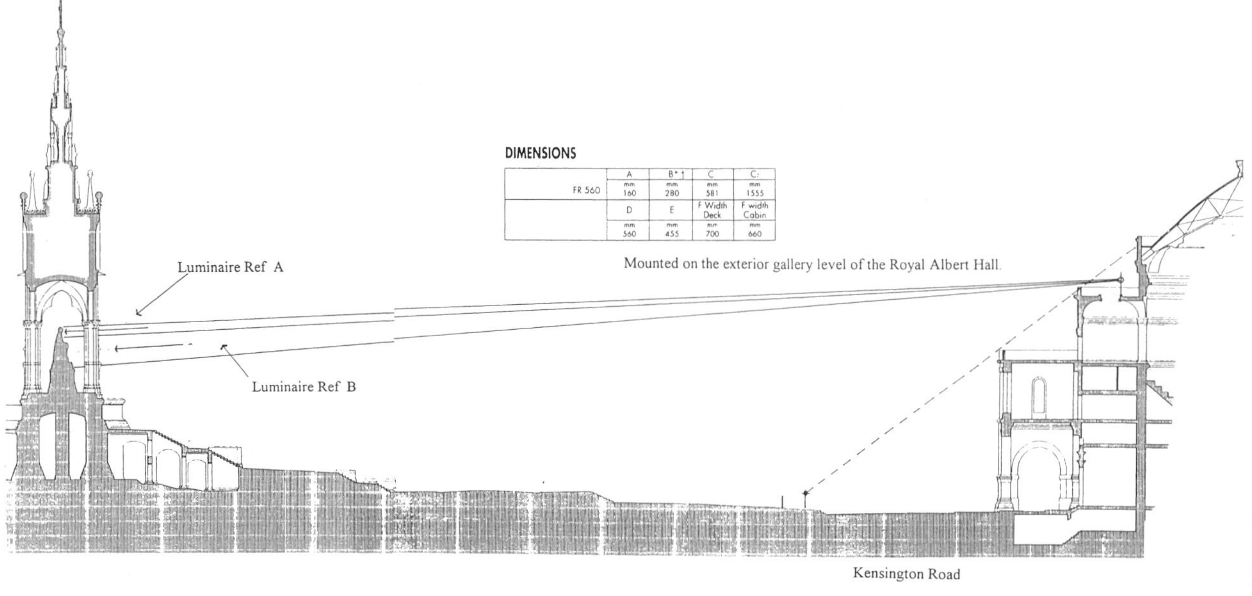

DIMENSIONS

	A	B* †	C	C:
FR 560	mm 160	mm 280	mm 581	mm 1555
	D	E	F Width Deck	F width Cabin
	mm 560	mm 455	mm 700	mm 660

Albert Memorial, section to show beam angles

Courtesy of Phoenix/Light Matters

Lighting designer Light Matters

Albert Memorial, long distance view

Albert Memorial, detail of subsidiary sculpture

Albert Memorial, detailed view of statue

scheme; for it is the frontal view of the memorial from Kensington Gore that is all-important. The park is closed at dusk when the lighting is turned on, so that glare which might have been experienced from behind the statue, is not apparent.

The lighting comprises:

1. The statue itself is lit from powerful narrow beam searchlights mounted on the Albert Hall; these are focused on the Prince's head and whole body, ensuring sufficient emphasis seen in relation to its surroundings.
2. The underneath of the canopy is uplit from concealed spotlights within the canopy itself, ensuring a contrast against the decorative ceiling.
3. The outside of the canopy, with its gold façades and pinnacle, is gently lit to avoid overpowering the statue of the Prince below.
4. An important part of the whole scheme is the lighting of the base frieze below the seated figure, and this is provided by spotlights concealed in special housings located in the landscaping opposite the four corners of the memorial.
5. The eight sets of statue groups at the four corners of the memorial are highlighted by narrow beam floods located in the special housings, which are set below the statues.

The lighting scheme has a sophisticated control system to enable it to react to the time of day and the seasons. Cool and warm tones have been used on the two diagonal axes to create subtle seasonal changes, emphasizing the modelling of the statue, and at the same time creating a series of scene-sets reacting to the lighting needs at different times, thus conserving energy. The most important aspect is that the lighting design has managed to avoid glare to the surrounding scene, while allowing enjoyment of the close-up inspection of the statue. After midnight only the statue of the Prince under the canopy is lit, while cool and warm light can be used at different times of year.

The main lamp used for the project was a metal halide (CDM-T) discharge lamp to ensure constancy of colour with efficiency. The main luminaire used was a marine searchlight providing narrow beams of light from a distance. The success of the appearance of the statue at night and the subtle location of the lighting equipment seen by day is a tribute to the dedication of the design team, and the funding by English Heritage.

Place de Terreaux, Lyon, France

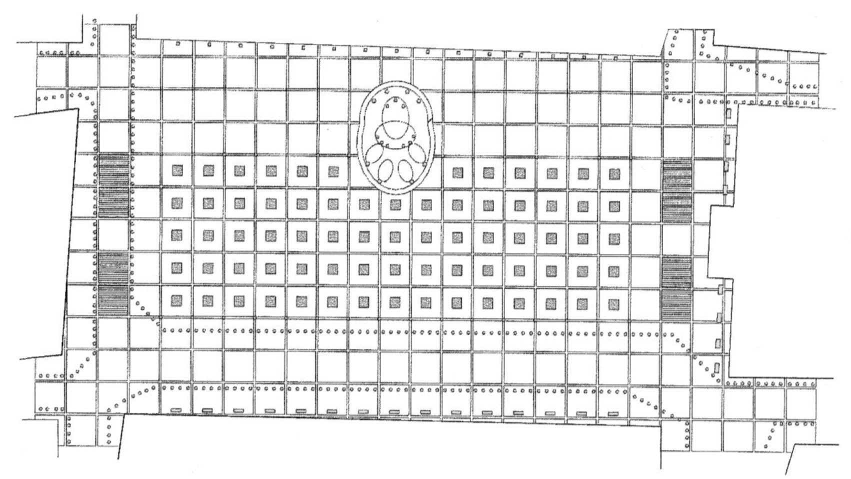

Place de Terreaux, plan of the square

Courtesy of Ville de Lyon

> As the French city of Lyon has recognized, lighting and nature need to be studied and thought out together. Light and snow, light and water, light and trees . . . light that enhances and highlights the beauty of nature, creating a wondrous magical quality.
>
> Stefano Marzano, Senior Director, Philips Design

The building of an underground carpark afforded an opportunity to redesign the 'Place de Terreaux' in the centre of Lyon. The concept for the redesign of the lighting was stated to be as follows:

> Dynamic and evolutionary lighting which follows the pace of activity in the city, its intensity varies according to the time of night . . . lighting which accentuates the contrast between the modern nature of the square and the ancestral architecture of its façades.

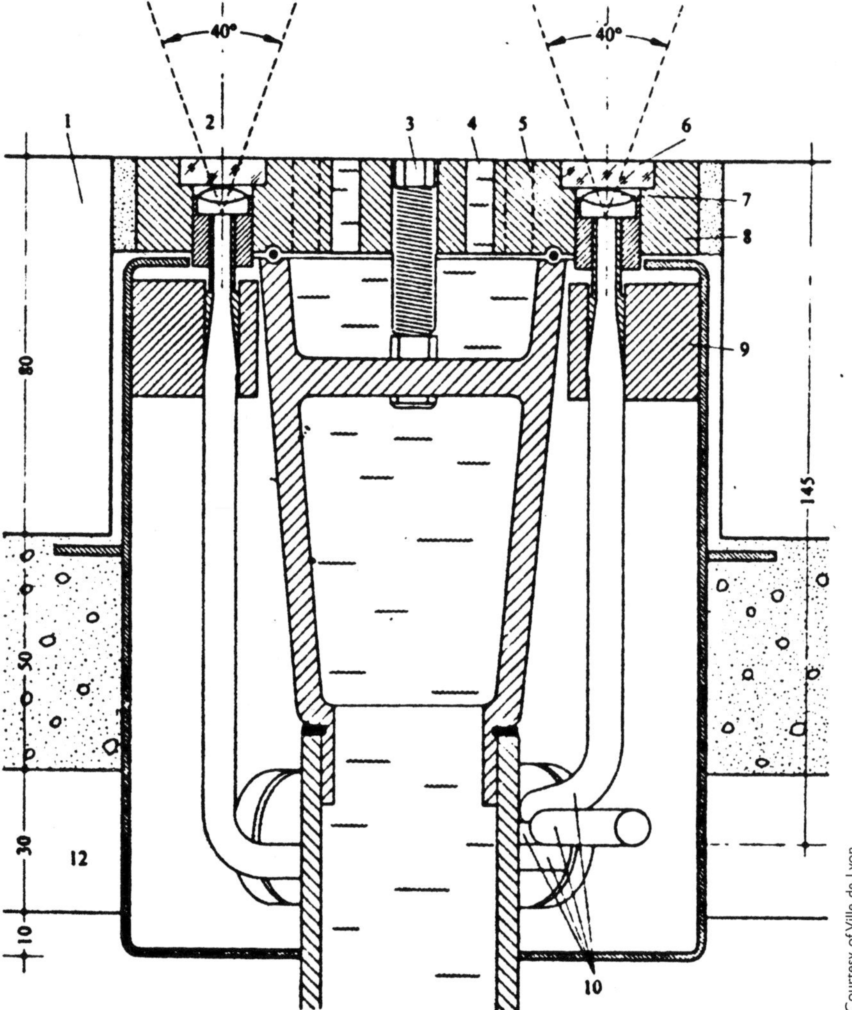

Place de Terreaux, detail of fountain lighting

Courtesy of Ville de Lyon

The undoubted success of the square at night is derived from a series of different elements adding up to a spectacular whole.

First, there is the exterior floodlighting of the historic buildings which define the four sides of the square, leading up to the Town Hall at one end. Three sides of the square are lit from ground level by means of recessed 2500 K (cool) high pressure sodium lamps integrated in a headlight with mirror and lens, producing a flat incident light to the façades. The lighting of the town hall is lit from building mounted floods concealed at cornice levels; these projectors light upwards to the decorative elements of the façade being designed to avoid both glare to the square, and light pollution to the sky.

Second, there is the combination of light with water, consisting of 69 mini-fountains recessed into the floor of the square, coupled with the splendour of the Bartoldi fountain.

The little illuminated fountains forming a dynamic pattern of light to guide people across the square can be controlled as to height (from 40 to 180 cm) and are lit by means of fibre optics using a cool (4000 K) metal halide source.

The Bartoldi fountain, on the other hand, is lit both from light sources concealed by rock formations in the basin, and from floodlights mounted on adjacent buildings. The fountain makes an emphatic incident in the square.

All lighting circuits are controlled by computer to enable the scene to react to the nature of the weather and the time of day. The design of the square and its lighting meets the needs of the population, and is robust enough to cater for the many uses of the square. For example, during 2000 the European Cup matches were displayed on a giant TV screen at one end of the square, which then accommodated several thousand spectators, with one side of the square lined with cafés – a memorable occasion.

Lighting designer Laurent Fachard (Les Eclairagistes Associes)

Courtesy of Christophe Lombard

Place de Terreaux, general view of the square (Architect Daniel Buren & Christian Crevet; Client Ville de Lyon, Antoine Bouchet & Jacques Fournier)

Courtesy of Louis Poulson)

Place de Terreaux, view close to the Bartoldi Statue (Architect Daniel Buren & Christian Crevet; Client Ville de Lyon, Antoine Bouchet & Jacques Fournier)

DP Archive

Place de Terreaux, view of the square during the European Cup (Architect Daniel Buren & Christian Crevet; Client Ville de Lyon, Antoine Bouchet & Jacques Fournier

Pirelli Gardens, Victoria and Albert Museum, London[1]

The lighting of the courtyard garden at the Victoria and Albert Museum, sponsored by Pirelli, was a response to the need for the museum to create additional income in order to make it more self-sufficient.

This was achieved by making the garden suitable for hiring-out for evening functions, corporate entertainment, and promotions. The way in which the lighting was designed is an example of the buildup of layers of lighting to achieve a final satisfactory appearance – a unified nighttime lit environment.

The illustration shows the court by daylight, and this is followed by the first layer of lighting to the door pillars and windows, lit by narrow

Pirelli Gardens, the gardens in daylight

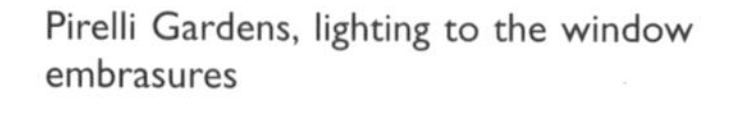

Pirelli Gardens, lighting to the window embrasures

[1]Phillips, D. (1997) *Lighting Historic Buildings*, p. 181. Architectural Press/ McGraw-Hill.

Lighting designer Lighting Design Partnership

beam spots, with additional fluorescent upward light to the tiled arches. The second layer comprises the lighting of the façade floodlit from adjacent buildings, using narrow beam spots lighting the pediment and roof statues. The spots are precisely focused to ensure that there is no glare to those in the garden. Finally, the trees are upward-lit from floods concealed in the landscaping around the trees, while additional narrow beam floods, mounted on the roof of an adjacent building light the fountain. When covered, the fountain can double as a podium for presentations, the lighting serving a double purpose.

In the context of an historic building, one of the design aims was first to ensure that any equipment mounted on the structure would be both safe and unobtrusive, and that the whole environment would be glare-free.

DP Archive

Pirelli Gardens, added frontal light to the façade

DP Archive

Pirelli Gardens, combined building with tree lighting

The Rows, Chester

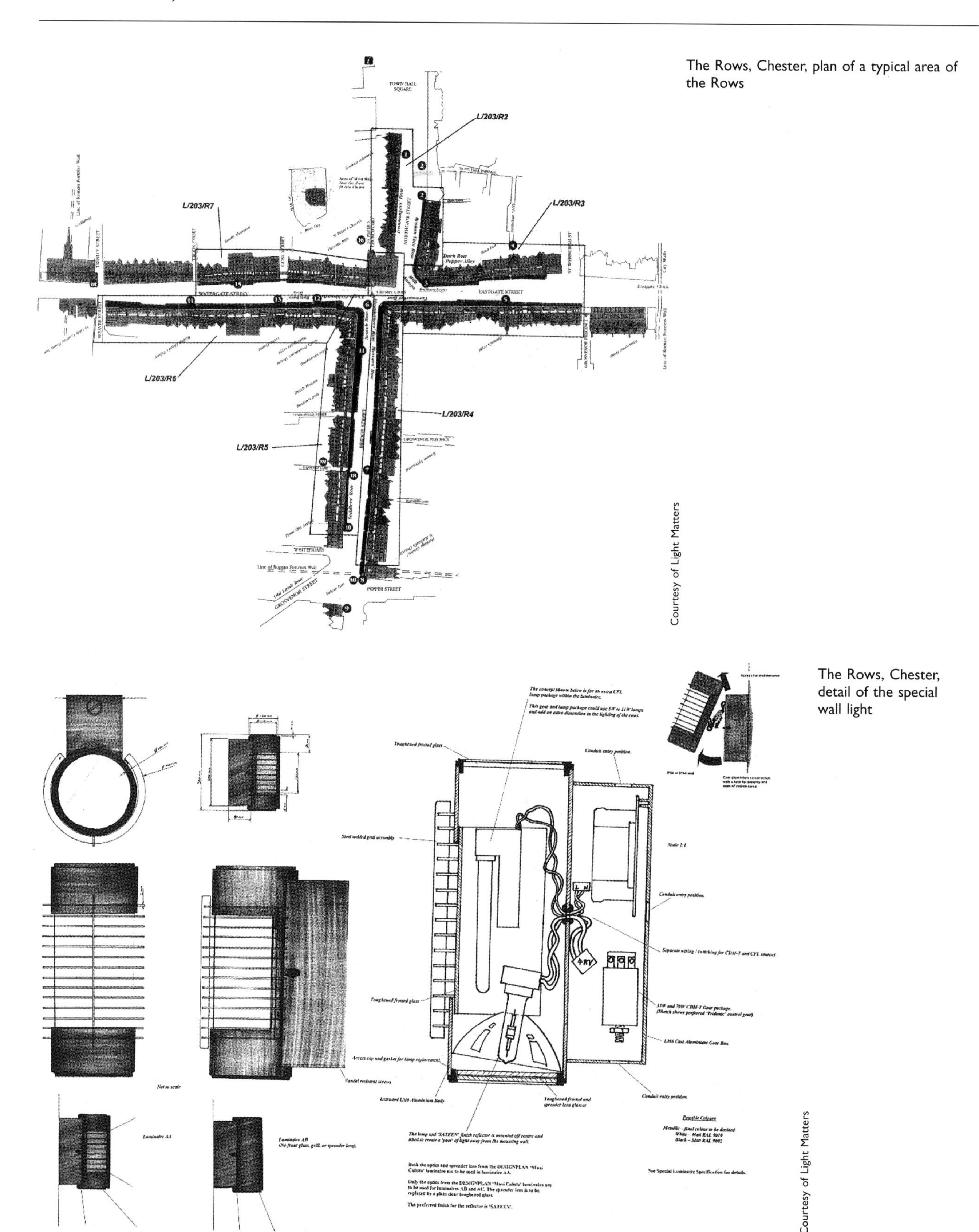

The Rows, Chester, plan of a typical area of the Rows

Courtesy of Light Matters

The Rows, Chester, detail of the special wall light

Courtesy of Light Matters

Lighting designer Neil Skinner, Light Matters

The Chester 'Rows' are a unique system of walkways which allow shoppers and others to do their shopping both at street level and at high level away from the street traffic. They were built in the Mediaeval period, and have survived the redevelopment of the town over the centuries.

They have always been one of Chester's great historic attractions and are as relevant today as in the past, combining the attractions of the 'high street' with the protection of the covered arcade.

The lighting scheme for the Rows was a part of the overall Chester Lighting strategy, having as its aims, amongst others:

1. Enhancement to the nighttime appearance of the town
2. Emphasis on lighting for people rather than for traffic
3. Reducing the dominance of highway lighting
4. To make the environment safer and deterring vandalism
5. Using energy efficient sources, reducing light pollution.
6. Careful attention to the daytime appearance of equipment.

Despite the historic nature of the area, an early decision was made to use modern light sources, and, where necessary, to design new light fittings to solve the problems involved. For this reason, a new 'Chester Lantern' was conceived by the lighting consultants in conjunction with Donald Insall Associates, the architectural consultants to the town, and this wall bracket fitting has been extensively used to provide the main lighting element throughout the scheme. The fitting contains two lamps: a 35 watt HQI 3000 K lamp shines downward from the fitting to illuminate pools of light to the walkways whilst a second 13 watt 2700 K compact fluorescent lamp provides the horizontal light through the centre grill and upwards to the ceiling. The whole design is unobtrusive and provides the lighting required, paying no undue lip service to the past.

Two other types of light fitting are also used. A 70 watt HQI 3000 K lamp is used to project a wide beam of light down towards the narrow staircases, highlighting the entrances to the Rows, helping to give a welcome to the views from the street. Finally, 50 watt 24 degree beam low voltage MR 16 halogen lamps are used to pick out decorative elements of the historic interiors.

The previous lighting to the Rows had left dark and forbidding areas, where residents and visitors seldom went. Now the new lighting scheme makes views from the exterior welcoming, and once in the Rows themselves makes walking and shopping a pleasure.

Courtesy of John Mills Group

The Rows, Chester, view to entrance stairs (Client Chester City Council)

Courtesy of John Mills Group

The Rows, Chester, view down rows/decorative spots (Client Chester City Council)

Courtesy of John Mills Group

The Rows, Chester, view of two-storey shopping arcade (Client Chester City Council)

Bond Court Gardens, Leeds

The Bond Court in Leeds may not rival the historic squares of some major cities, but it shows what can be achieved with modest means to improve an unattractive commercial area.

Bond Court was a large barren square surrounded by modern high rise offices in the city's business district, used mainly as a car park for vehicles requiring access, where the public were not encouraged to linger. Despite this, some 3000 pedestrians per hour would pass through the square during peak lunchtime hours – 'pass through' being the operative words.

A trust was set up to honour the memory of one of Leeds' senior citizens, with the aim of improving the city, making it greener and giving it a continental feel, in line with Leeds' aspiration to become a major European City. This fund, set up by The Scurrah Wainright Charitable Trust, paid for the work.

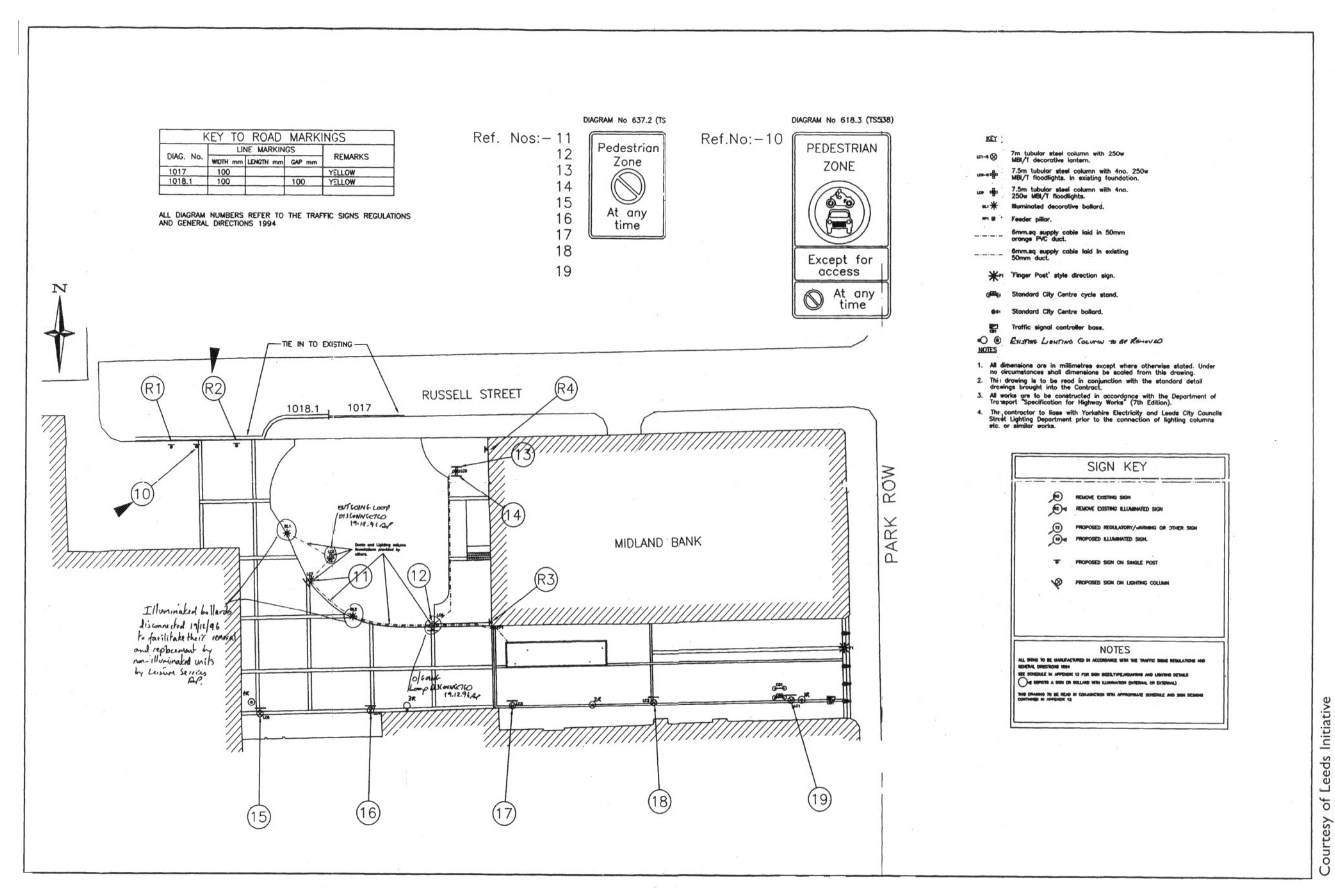

Bond Court Gardens, Leeds, plan of the square indicating the lighting

Lighting designer Leeds Initiative

The result is a small oasis for office workers and others, and includes a boules court and chess tables, which have proved a popular attraction during lunchtime hours. The area has been transformed, and has had a spin-off by initiating improvements in neighbouring areas. Particular attention has been paid to the landscaping, the surfacing materials and the lighting: while simple pole mounted lanterns are used, being designed for safety and security, they give sufficient nighttime lighting for the boules court; indeed this may even have the effect of introducing this charming continental pastime to this country.

It is in such modest ways that areas may be transformed, and our cities renewed.

Courtesy of Leeds Initiative

Bond Court Gardens, Leeds, court from high level, night

Arundel Town Centre

Inconspicuous, building mounted floodlights, painted black or white to match buildings, 11,000 lumen (150w) 3K metal halide lamp. (Kim)

Traditional 4m cast iron columns topped with lantern fitted with road lighting optic and 6,000 lumen (70w) 3K metal halide lamp. (Sugg)

Cast iron wall bracket attached to castle walls with lantern fitted with road lighting optic and 6,000 lumen (70w) 3K metal halide lamp. (Sugg)

Courtesy of Nigel Pollard (NEP)

Arundel Town Centre, plan of the lighting

The centre of Arundel had in the past been poorly lit, with old sodium (SOX) lamps, giving patchy light of an unpleasant colour, with out-of-scale road lighting lanterns and concrete columns.

In a gentle scheme of road and amenity lighting, instigated by the Arundel Society, the area has been transformed, while recreating the quality associated with an old historic town. The light sources used throughout are the 70 watt and 150 watt metal Halide (CDM) lamp. This provides a warm (3000 K) colour, giving a welcoming appearance to the town centre.

Wall-mounted fittings are of two main types: inconspicuous building mounted floodlights, at high level painted to fit in with their building background, black or white, and cast iron wall brackets attached to the crenellations of the walls to the grounds of Arundel Castle (left side).

All the old 8 metre high 1950s concrete columns were replaced, wherever possible, by building mounted units and where this was not possible, or extra light was required, 4 metre cast iron 'Arundel style' heritage columns were installed. These, using the same metal halide lamps in modern optics, helped to provide pools of pedestrian lighting, with a vertical focus where necessary.

The scheme might well be described as a heritage solution, but using the most up-to-date lighting technology to achieve a worthy lit environment for this historic town.

Lighting designer Nigel Pollard

Courtesy of David Nicholls

Arundel Town Centre, longer view, from the castle wall

Somerset House, London

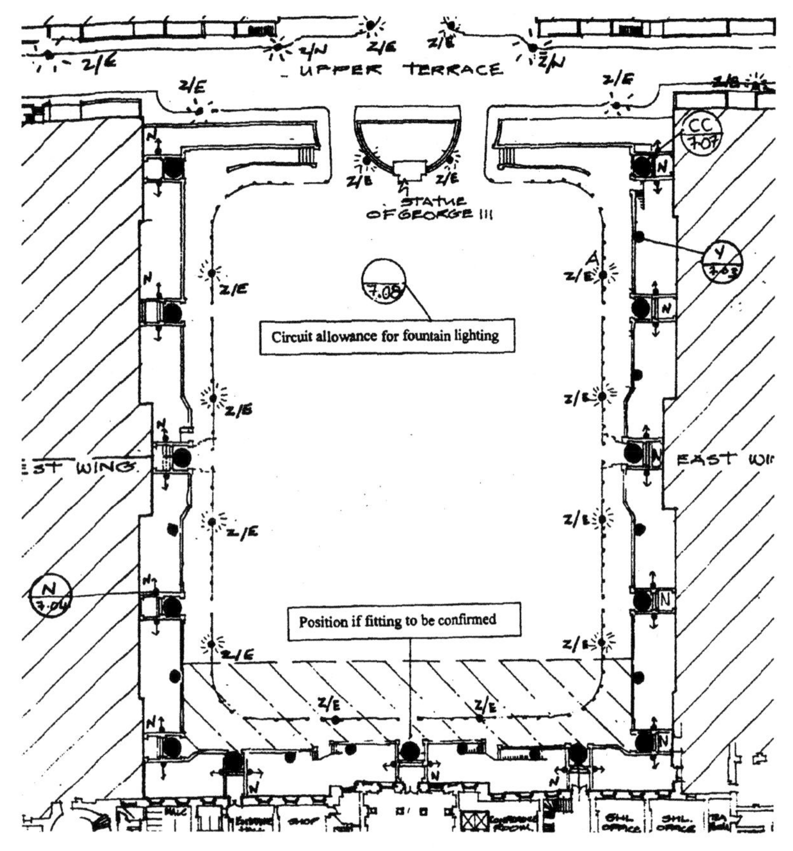

Z/E EXISTING LAMP STANDARDS
Z/N NEW LAMP STANDARDS

THE GREAT COURT, SOMERSET HOUSE

Somerset House, plan of the square, lighting layout

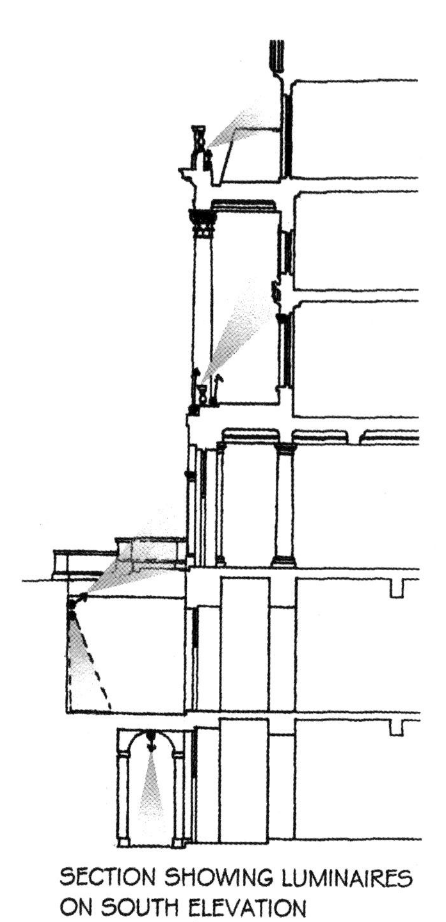

Somerset House, section through side buildings/floodlights

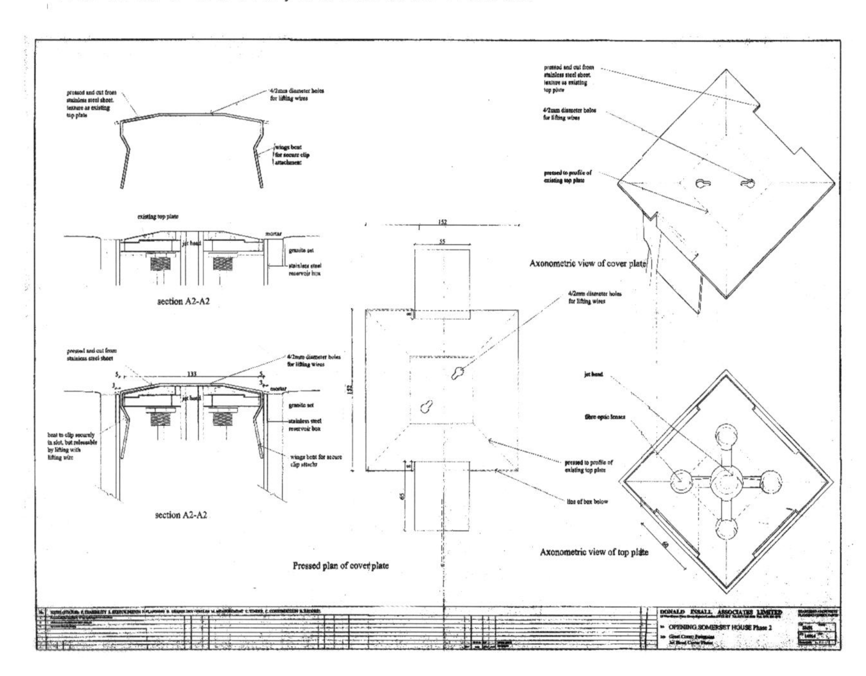

Somerset House, water jets with fibre optic heads

Lighting designer Sharon Stammers of Light Matters

Somerset House was originally built as Government offices, completed by Sir William Chambers in 1856. It was best known until recently for the fact that it was where the UK kept is records of births, deaths and marriages, while the fine central courtyard had been wasted as a staff car park.

The buildings themselves now offer an important cultural complex, containing the Courtauld Gallery, the UK extension of the Hermitage Gallery, the Gilbert gold and silver collection and the Admiralty restaurant. The central courtyard with its surrounding buildings has been transformed and the resulting open space developed for public use during the day. At night it can be hired for corporate entertainment. This space has all the elements of a lit environment at night. It consists of the floodlit buildings surrounding the courtyard, with incidents such as the illuminated statue, and the 'Edmond J. Safra fountain' as dynamic attractions in its centre.

The buildings are lit from below, having the advantage of a sunken area in front of the building below ground floor level; where the asymmetric wide-beam floodlights may be concealed (see right side of illustration on opposite page) the use of the metal halide (CDM) lamp ensuring colour consistency, with energy efficiency.

The large statue of George III at one end of the square acts as a focus, but it is the Safra fountain where 55 water jets spring from a pattern of stones in the courtyard which provides the magic during the day and night, while the water jets can be varied in turn as to their height, up to a maximum of 5 metres.

Each fountain is lit by means of a sophisticated system of fibre optics, which can be varied in colour. Each jet has four fibre optic heads, lit by means of 22 concealed light boxes containing 150 watt HQI/R lamps. Two colour wheels are integrated into each of the lamp boxes, these overlap to provide 36 different colours, in such a manner that a symphony of colour can be orchestrated. Bearing in mind the flexibility of height and hue, an almost limitless light show can be produced.

A further dimension has been added, the ability to create an ice rink, and over Christmas 2000, this attraction was provided for the enjoyment of the public in the centre of the square.

It is of interest that the Safra fountain in the courtyard at Somerset House is the first major public fountain since those designed by Sir Edwin Lutyens in Trafalgar Square in 1939, also included as a Lit environment (see p. 131) illustration. It is to be hoped that it will not be the last.

Courtesy of Nick Wood

Somerset House, daytime in the square (Architect Donald Insall)

Courtesy of Peter Durant

Somerset House, night view courtyard/coloured fountains (Architect Donald Insall)

Courtesy of Nick Wood

Somerset House, night view/ice rink (Architect Donald Insall)

Boulevard Richard Lenoir, Paris

The development project for the pedestrian walkway in the wide central reservation of the boulevard was initiated by the City of Paris in 1992 and completed in 1994. The project which continues for 2 kilometres, consists of two parallel walkways, one on either side of a series of gardens and spaces for open markets held each week.

The Boulevard itself is of great historic interest in that it is located above the old St Martin's canal which leads from the Seine in central Paris to the Bastille. This helps to give the boulevard its essential character, due to the 37 three metre circular openings along its length, which provide the canal below with light and air. These openings or 'oculi' are protected by a mesh for safety reasons, but otherwise are open to the sky – they are lighted from below, glowing with light at night.

The lighting of the pathways is achieved by attaching two lines of 'Lyre Candelabras' to the side row of trees; but the most spectacular incident along its length consists of 18 groups of vertical water jets lit by fibre optics, and designed to reach up to a height of 2 metres. Each water jet is incorporated within the pavement pattern having an associated fibre optic head (see illustration detail below).

Each of the groups of jets contains two parallel rows of seven jets each. The view of the fountains leading up to the Bastille tower shows the visual impression.

The total effect of the Boulevard Richard Lenoir is derived from being combined with the canal St Martin below, well known to the boating fraternity, and the pedestrian reservation with its lighted incidents above.

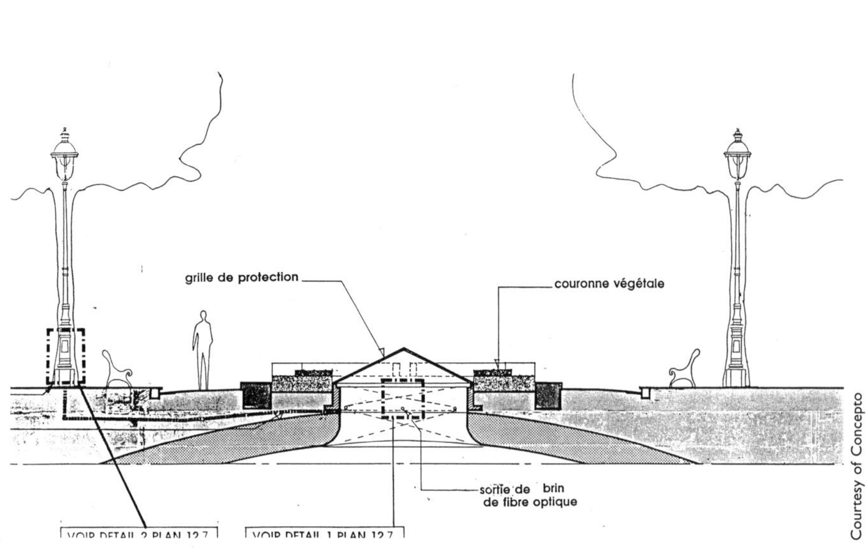

Blvd Richard Lenoir, section to show 'Oculus' relationship length

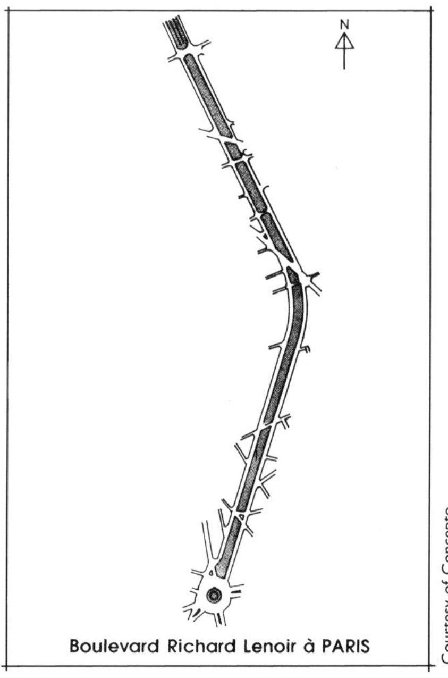

Blvd Richard Lenoir, plan of 2 kilometre length

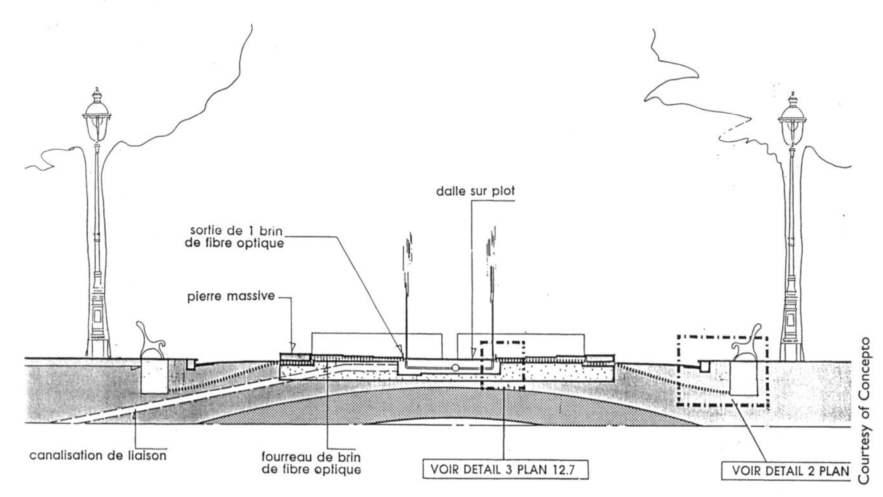

Blvd Richard Lenoir, similar section to show fountains

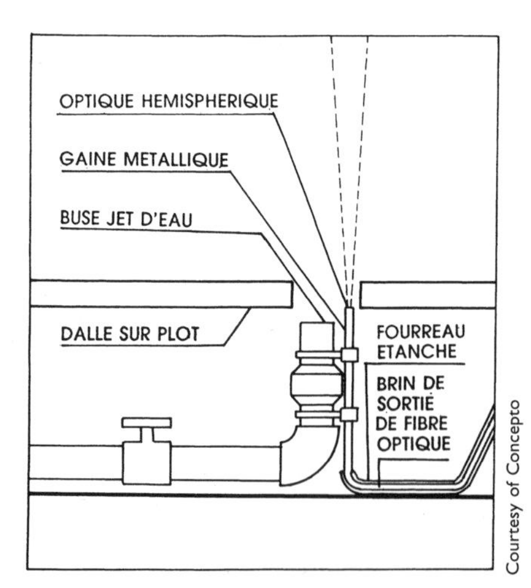

Blvd Richard Lenoir, water jet and fibre optics detail

Lighting designer Roger Narboni, Concepto

Courtesy of Concepto

Blvd Richard Lenoir Oculi by pedestrian walkway (Architect David Mangin, Seura Agency; Landscape Architect Jacqueline Osty)

Courtesy of JP Cousin

Blvd Richard Lenoir, view toward 'Bastille' column (Architect David Mangin, Seura Agency; Landscape Architect Jacqueline Osty)

Courtesy of Concepto

Blvd Richard Lenoir pathway and central reservation (Architect David Mangin, Seura Agency; Landscape Architect Jacqueline Osty)

Trafalgar Square, London

Trafalgar Square, aerial view

Courtesy of Donald W. Insall and Associates

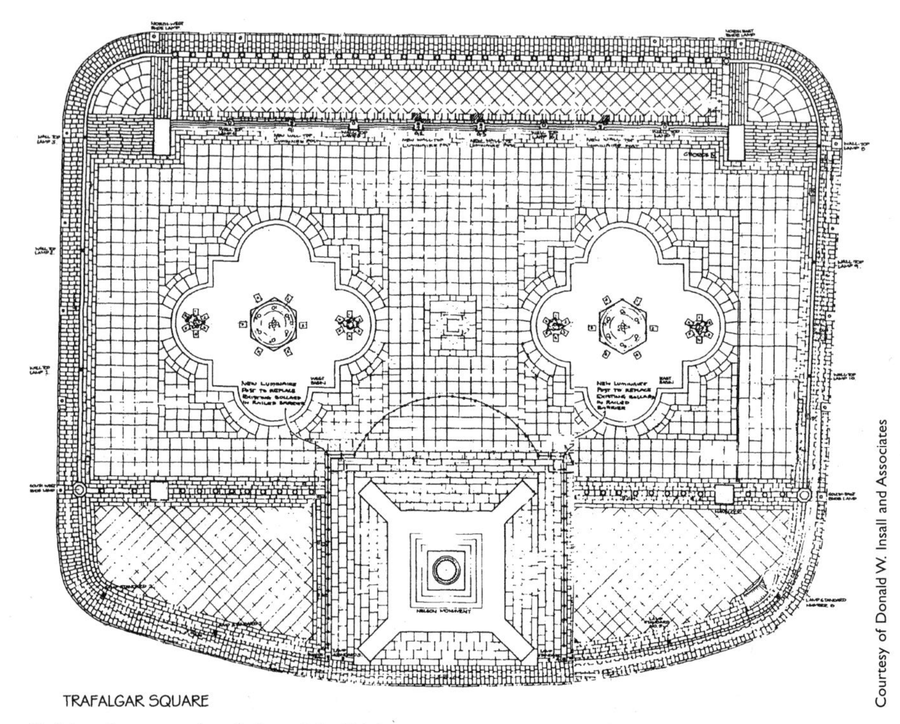

Trafalgar Square, reduced plan of the lighting

Courtesy of Donald W. Insall and Associates

Following a report on the square by the Department of National Heritage in 1994, the architects Donald Insall Associates who had previously designed the refurbishment of the paving, were appointed as consultants with Lighting Design Partnership to redesign the lighting. The scheme, which was commissioned by Westminster City Council, was envisaged as a master plan taking into account the surrounding buildings. Visualizations of the square and surrounding buildings have been included in Chapter 1 on Strategy (p. 13) and it will be useful to compare these with the situation in the square today.

The lighting proposals can be summarized as follows:

1. The lighting of Nelson's Column
2. The lighting of the Lutyens fountains
3. The lighting of the National Gallery and surrounding buildings
4. Improved amenity lighting.

The most difficult lighting problem was that of Nelson's Column, with the statue of the famous Admiral at the top. The statue is lit from narrow beam 400 watt metal halide lamps mounted at the top of adjacent buildings. The column, base and Landseers lions are lit by 1000 watt CSI fittings, using PAR 64 and CDM-T lamps. The fittings are concealed in new cast iron bollards, specially designed to blend with the layout and form of the existing historic fabric.

The fountain lighting is integrated with the Lutyens structures, using low voltage tungsten halogen lamps to ensure electrical safety. The wider, lower, middle and upper basins have underwater fittings.

To improve the overall amenity lighting of the square, the existing cast iron columns were increased in number, the optics being changed to reduce glare. In general, metal halide lamps have been used, rather than the warmer HP sodium (SON) used for street lighting locally. The purpose was to unify the lighting of the square and surrounding buildings, including the whole length of the National Gallery; very much a part of the lighting of the square. See Lighting Bollards, pp. 99 and 101.

Lighting designer Lighting Design Partnership

DP Archive

Trafalgar Square, lighting to Nelson's Column (Architect Donald Insall Associates)

DP Archive

Trafalgar Square, lighting to National Gallery and fountain (Architect Donald Insall Associates)

DP Archive

Trafalgar Square, lighting to Lutyens Fountains with St Martin in the background (Architect Donald Insall Associates)

DP Archive

Trafalgar Square, day view of underwater light to fountain (Architect Donald Insall Associates)

Bristol regeneration

A map of the centre of Bristol illustrates the many open spaces which have been provided. The two which are featured here are:

1. The Millennium Square
2. The associated Anchor Square.

These are both a part of the regeneration of Bristol's dockland area, which has been taking place over the last decade. These two interlinked squares relate and provide access to two important buildings: the 'Explore Bristol' museum and the 'Wildscreen'. A lighting brief was prepared by Alec French Partners of the Concept Planning Group; it is of interest to quote from this, as it shows the importance which was given to the subject, and which contributed to the success of the project.

Millennium Square

The square is built over two floors of car parking, and this provided a lighting opportunity with the series of ventilation towers which line the square, featured with blue light. It is these, together with the Explore Bristol building on the one side which define the edges of the square. No attempt has been made to provide bland overall lighting to the square, dearly loved by local authority engineers. Lighting is derived from the incidents, water features and lit landscape and trees in the square.

An important incident in the square is the 'Imaginarium' silver sphere, which dominates the square on the side of the Explore Bristol building. Of particular interest is the figure of eight, or 'Analemma' by artist David Ward formed in the paving of the square by means of 52 'helicopter pad landing lights'. These lights play all sorts of games, by a changing sequence of colour. This has caught the imagination of the public and in the summer children play in the fountains and chase the lights.

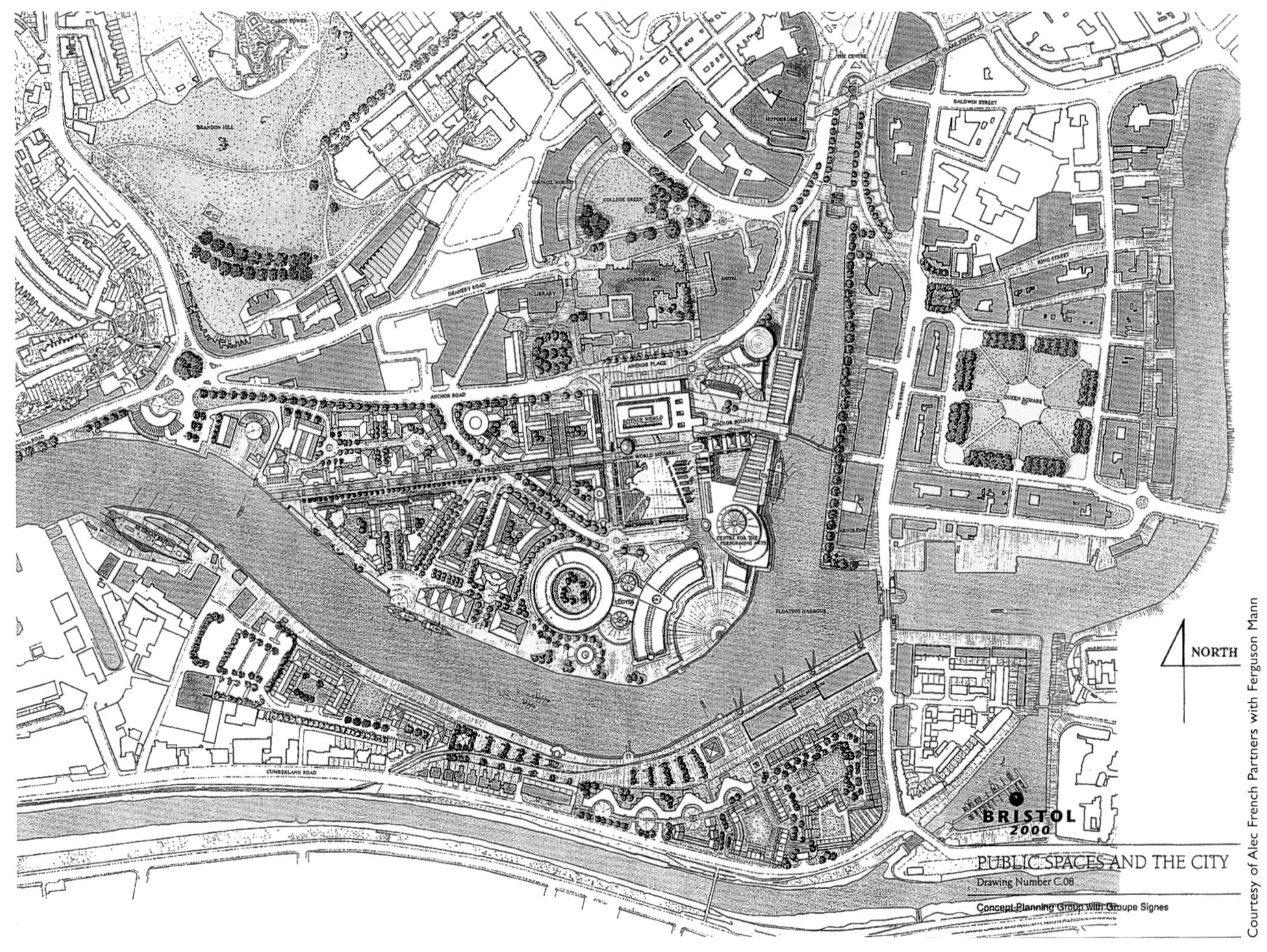

Courtesy of Alec French Partners with Ferguson Mann

Bristol Regeneration, town plan showing public spaces

Lighting designer Alec French Partners with Ferguson Mann

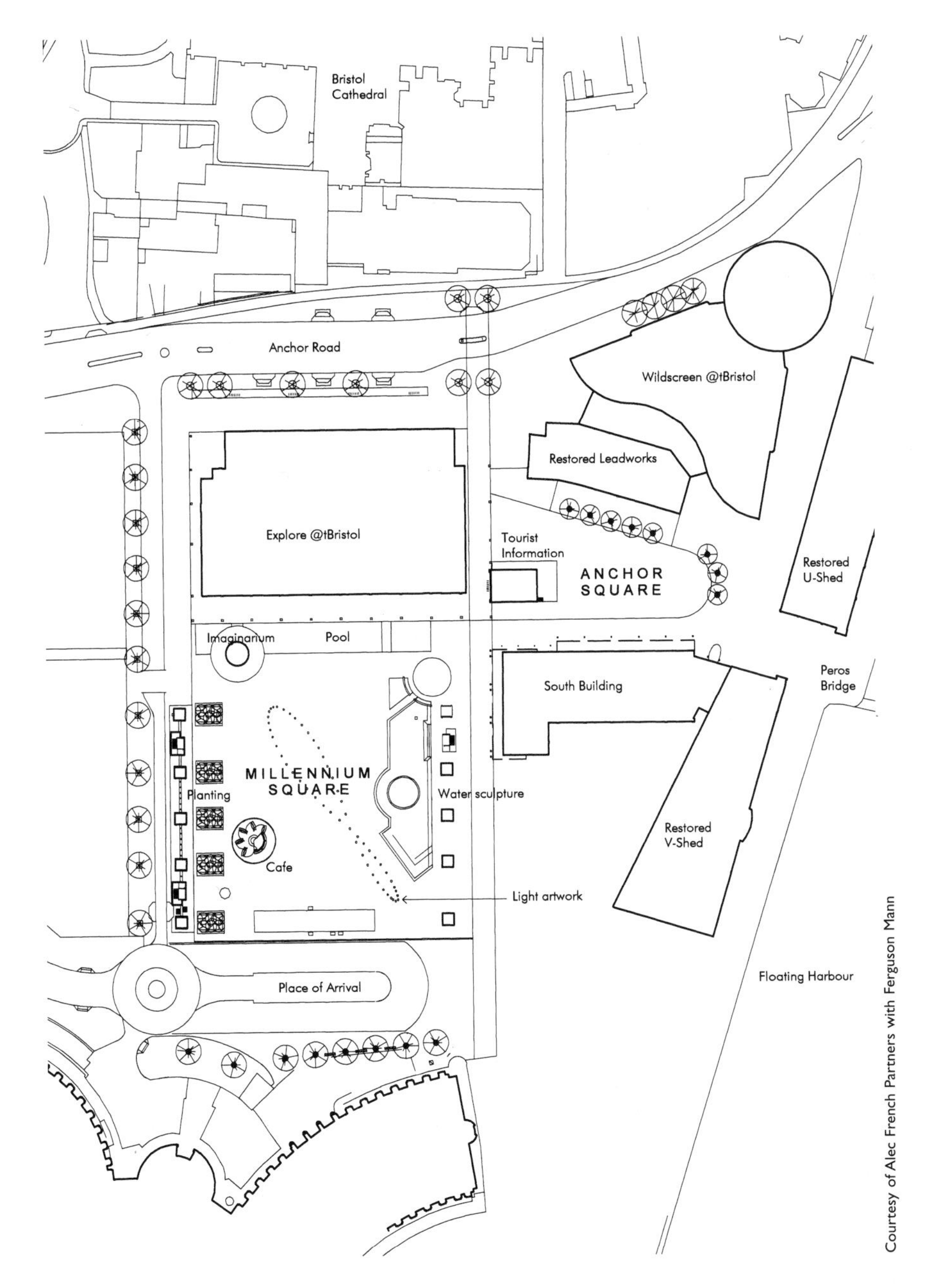

Bristol Regeneration, plan of Millennium and Anchor Squares

Anchor Square

This is an interim space giving access to both buildings, the restored leadworks and on towards Millennium Square. As such, it is lit both from the light of the adjacent buildings, and from a series of well-designed pedestrian pole lights and recessed tree up-lighting. Apart from a delightful sculpture of a scarab beetle by the artist Nichola Hicks, which remains unlit at night, it lacks lighting opportunities in the form of incidents, and provides a gentle anti-space to the excitement of Millennium Square.

Bristol regeneration (*continued*)

DP Archive

Bristol Regeneration, view of Anchor Square lighting (Architect Alec French Partners with Ferguson Mann; Landscape architect Balston & Co)

DP Archive

Bristol Regeneration, view of Millennium Square towards silver sphere (Architect Alec French Partners with Ferguson Mann; Landscape architect Balston & Co)

Lighting designer Alec French Partners with Ferguson Mann

DP Archive

Bristol Regeneration, night view across the square (Architect Alec French Partners with Ferguson Mann; Landscape architect Balston & Co)

Lighting brief

1. The evening and nighttime use of the public spaces should provide an attractive, enjoyable, occasionally dramatic, and always safe, environment for the users. Lighting design should address all these requirements and in ways which complement the other features of the spaces. Lighting will be integrated within structures where appropriate, with light sources being discreet. Freestanding light fittings will be chosen to relate well to other items such as seats, bollards and litter bins
2. Coordination of the lighting of individual buildings with the lighting of the spaces will be important
3. Floodlighting should be considered for particular features, e.g. restored leadworks chimney and freestanding works of art
4. Proposed 'light artists' work will be fully integrated with the fabric of the spaces.

Courtesy of Adam Wilson

Bristol Regeneration, night view of the Analemma (Architect Alec French Partners with Ferguson Mann; Landscape architect Balston & Co)

University campuses

Jubilee Campus, Nottingham University

The Jubilee Campus is an expansion of Nottingham University, to house three faculties as well as a central teaching facility, a Learning Resource Centre, a central dining facility, and 750 rooms within three halls of residence. The campus is basically a linear development planned alongside a lake which almost encircles a building housing the library (see plan).

The exterior lighting consists of a few basic, but well-conceived elements, to which the architects had a major design input.

The elements are as follows:

1. A lighting column, consisting of an upward light source concealed in the base of the fitting, lighting up to a reflective plate which distributes a gentle light downwards. The fitting is used both internally and externally and can be seen in the photographs of the café (internal) and the promenade (external).
2. A special lit bollard located at positions in the lake to give directions to the different buildings. These are a major decorative feature in the lake, while having an essential functional use. The bollards can be seen in the lake alongside the promenade.
3. Low-level bollards give safe access to the walkway along the lakeside, but where the main buildings align the lake these have been omitted as unnecessary, since the light from the buildings is sufficient.

The lake provides a reflective surface for the lighting, with images of the linear buildings seen from the promenade.

Jubilee Campus, Nottingham University, sketch plan of the overall layout

Courtesy of Michael Hopkins

Lighting designer Ove Arup and Partners

Jubilee Campus, Nottingham, view across the lake to the library (Architect Michael Hopkins)

Jubilee Campus, Nottingham, promenade with main buildings on right (Architect Michael Hopkins)

St John's College, Oxford

The exterior lighting of the garden quadrangle at St John's College, Oxford, is designed to reinforce the contrasting character of the spaces above and below the terrace level – in the words of the architect: 'The ground floor of the building was conceived as an underworld, with domes over the main spaces and a large garden terrace above with views out over the college grounds.'

The central circular court is open to the sky and the exciting visual effect of this is apparent in the first photograph, showing the garden at the top, with the underworld below. The other two spaces at ground level are enclosed by domed ceilings, one being used as an auditorium and the other as a dining-room. The lantern-like cupolas above each of these spaces make a significant contribution to the nightscape of the garden quadrangle above. The terrace (at first floor level) over which all the study bedrooms look, is lit both to provide a safe environment and, in addition, to light the surrounding garden layout and planters, to achieve a restful visual appearance, by means of lights recessed into the low side walls, with the planted areas illuminated by spill light from the windows of the study bedrooms. By contrast, the white precast concrete structure of the central court, at ground level, is up-lit by lights set into the paving, below chains taking the terrace rainwater.

Courtesy of McCormac Jamieson Prichard

St John's College, Oxford, section and ground level plan

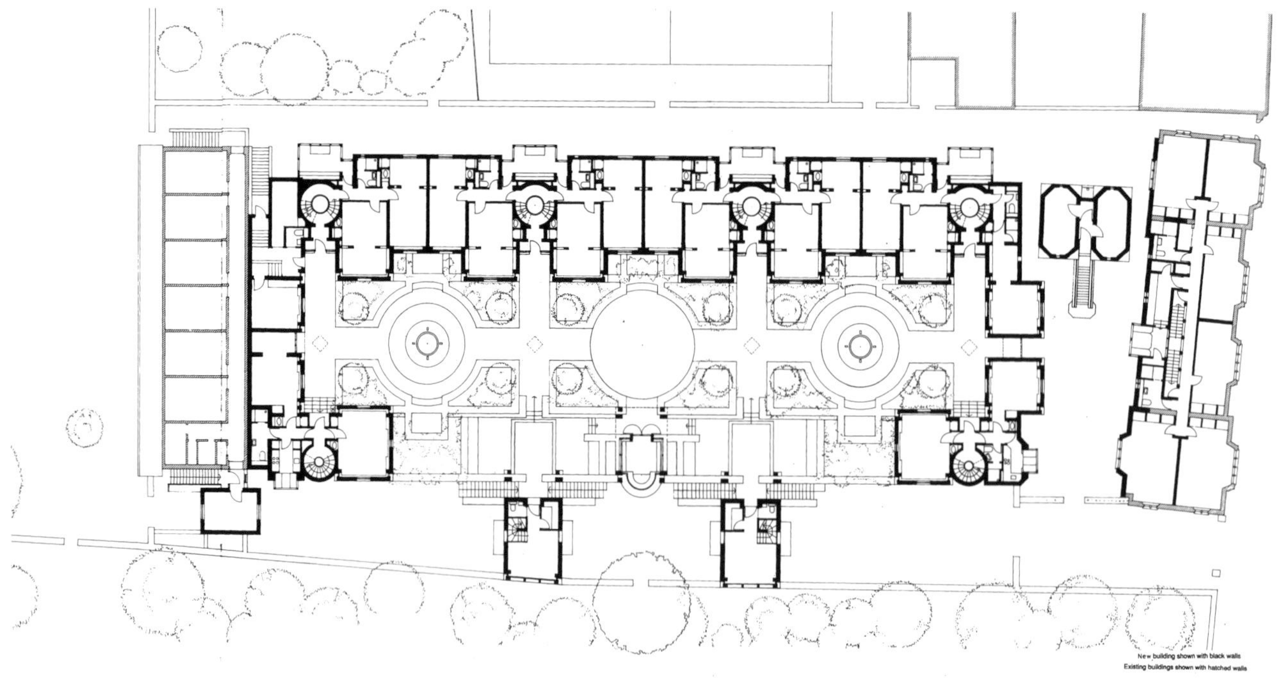

ST JOHN'S COLLEGE OXFORD RURAL ECONOMY SITE TERRACE LEVEL 1:100 AUGUST 1990

Courtesy of McCormac Jamieson Prichard

St John's College, the garden quadrangle (Architect McCormac Jamieson Prichard)

Lighting designer Steenson Varming and Mulcahy

Courtesy of Peter Durant

Courtesy of Peter Durant

St John's College, view of the undercroft (Architect McCormac Jamieson Prichard)

Commercial/shopping/offices

Bluewater, Kent

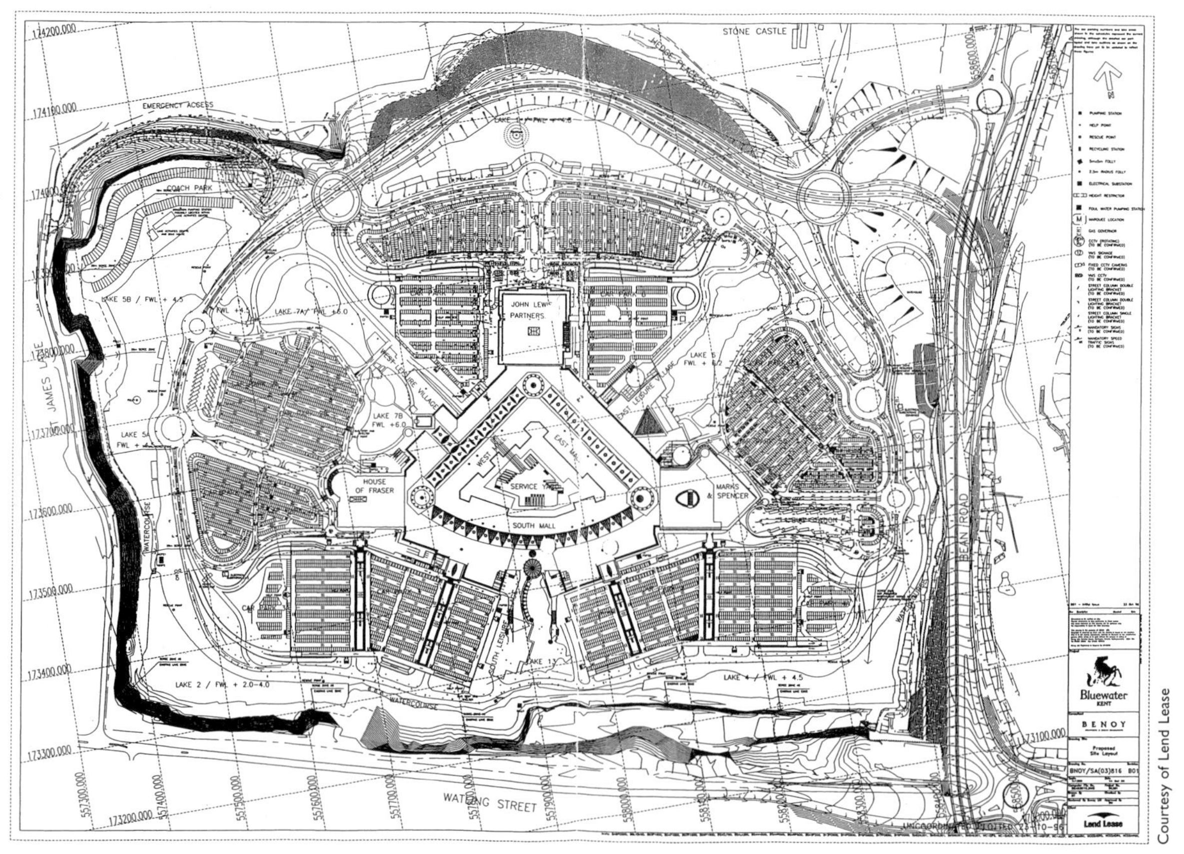

Courtesy of Lend Lease

Bluewater, plan of the complex

DP Archive

Bluewater, view of building reflections and lakes (Architect Eric Kuhne Associates; Benoy; Landscape architects Robert Townshend; Client Lend Lease)

Lighting designer Speirs and Major

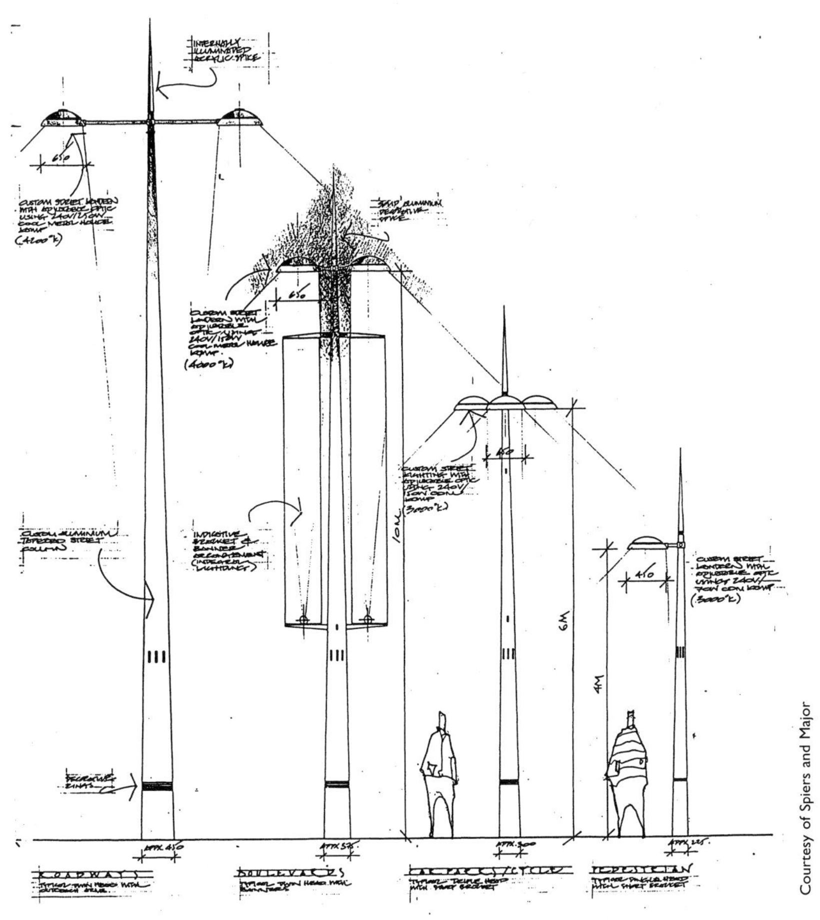

Courtesy of Spiers and Major

Bluewater, family of lighting columns

Courtesy of Spiers and Major

Bluewater, trees and pathway (Architect Eric Kuhne Associates; Benoy; Landscape architects Robert Townshend; Client Lend Lease)

In order to understand the exterior lighting at one of the largest shopping complexes in Europe, it is important to break it down into component parts, while at the same time realizing that it is all a part of one lit environment. The simple breakdown is as follows:

- Car parking and roads
- Trees and pedestrian areas
- Lakes and landscaping, their relationship with building façades.

Car parking and roads
On-site car parking consists of both 'decked car parks' and 'open parking spaces'. The tops of the decked parking spaces are further developed as open parking spaces. The lighting of the decked areas was achieved with a polycarbonate 'biscuit' close up to the ceiling, containing two 40 watt CFC lamps. To give some idea of the size of the project 8500 fittings were required. The open areas were achieved with a special design of post top lantern, as a part of a family of fittings developed for the project, giving lighting also to roads and pedestrian areas. The associated sketch indicates the design of the family of fittings, with a characteristic 'blue' dome on the top. The design of lantern with a horizontal cut-off reduces as far as possible the effect of light pollution from the fitting; while it can do nothing to reduce that due to the reflective capacity of the roads and car park areas.

Trees and pedestrian areas
The trees defining the pedestrian ways through the car parks are uplit with metal halide (CDM) fittings, allowing some spill light to the pathways. The fittings are fixed to specially designed concrete pads, and are fitted with louvres to minimize glare to passers-by. Where pathways are not bordered by trees, they are lit using lighting bollards.

Lakes and landscaping: their relationship with building façades
No lighting equipment is submerged in the lakes and water features. The reason for this was to ensure that the lakes have a natural appearance, and also for safety and ease of maintenance. The decorative lighting to the general soft and hard landscaped areas was created by a mix of small buried floods using LEDs to mark paths and define edges.

An important feature of the exterior nightscape is the relationship with the interior lighting of adjacent buildings, where the lighting from the interior spills out to the landscape. Water also plays an important part, by the interreflection of light from the buildings to the water and the exterior landscape.

In some areas exterior dynamic coloured sources are played on to the exterior surfaces of the buildings, and it is this mixture of dynamic and static elements, providing as it does an increased intensity close to the building, which contributes to the overall success of the project.

Hamilton Park, Glasgow

Focal Point

Grand Avenue

Fence

● PHILIPS RECESSED KBF505W55
○ PHILIPS CGP440/02/ZPS441/1x CDM-T 70W
CONTROL PILLAR

Courtesy of Philips Lighting

Hamilton Park, plan of the Grand Avenue, lighting layout

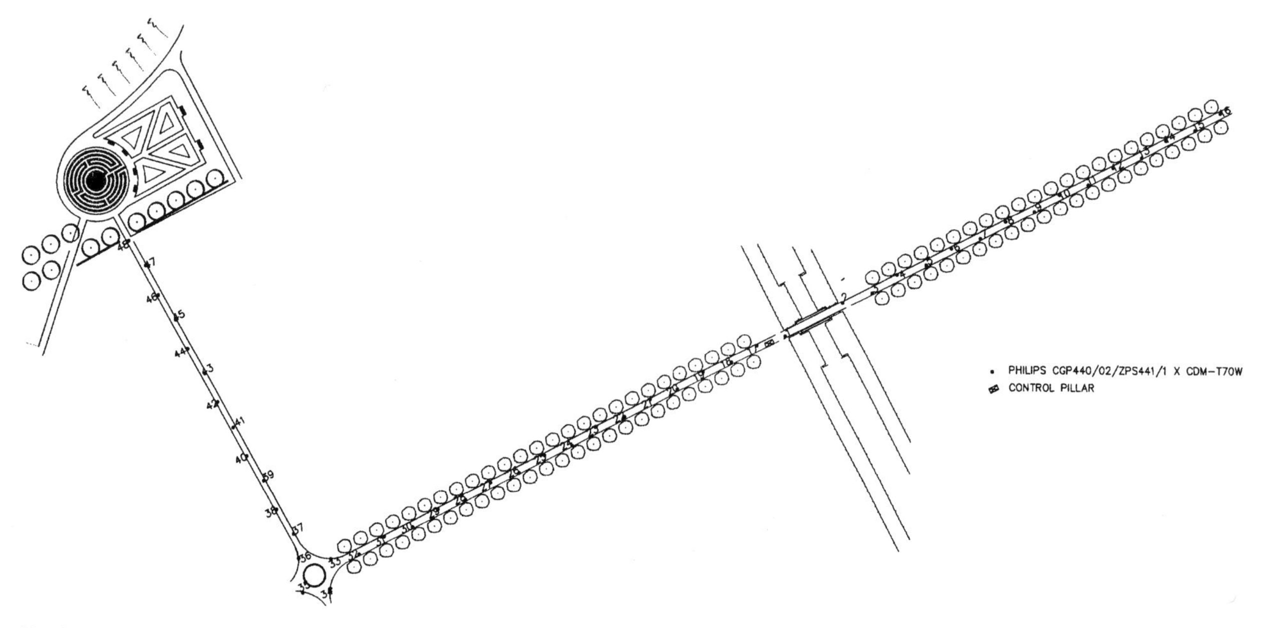

Courtesy of Philips Lighting

Hamilton Park, plan of a minor avenue leading up to the Mausoleum

Lighting designer Philips Lighting

Courtesy of Philips Lighting

Hamilton Park, the Grand Avenue shows special column design (Client South Lanarkshire Council)

Courtesy of Philips Lighting

Hamilton Park, the Mausoleum pathway, leading to floodlit building (Client South Lanarkshire Council)

The area, 15 miles south of Glasgow and once the home of the Dukes of Hamilton, but now owned by South Lanarkshire Council, the Hamilton Palace grounds is in the process of being developed for new sporting facilities and retail outlets.

At present the infrastructure has been completed, consisting of 5 miles of external pathways split into two major areas: the Grand Avenue and the Mausoleum Path.

Both pathways end up at the Mausoleum, but the Grand Avenue is the more important of the two, being 450 metres long and 30 metres wide. The avenue consists of two corridors of mature trees, one housing a footpath. A decision was made to uplight each of the 151 trees using a 55 watt QL light source recessed into the ground between the trees. The QL light source was chosen for low maintenance, low surface temperature, and excellent colour rendering of the source, despite its additional expense.

The lighting of the avenue and pathways employs a special 3.5 metre column, set at 15 metre centres with a remote light source set in the base lighting up to the reflector at the top, and at the same time lighting up polycarbonate strips set in the side of the column. The light source used is the 70 watt metal halide CDM-T source, giving consistent colour rendering with good energy management. Minor pathways are lit with the 35 watt metal halide CDM-T source, to ensure colour balance throughout the site.

The lighting columns are designed to light downwards to ensure minimum glare to the walkers, and virtually no light pollution to the night sky. A fundamental aspect of the design is to ensure that the public using the area will feel secure, and this has been achieved.

Yorkshire Building Society

The development of the site pays sympathetic regard to the neighbouring residential areas, and is designed primarily to provide a safe environment for vehicle access and staff security.

The exterior lighting is closely linked to the lighting of the building itself, which is of mixed form requiring a degree of flood lighting from external sources. The pattern of windows provides identification of the building form for all but the main entrance, which has its own lit canopy.

The lighting of the brick façades is provided by 70 watt SON-T lamps incorporated into the eaves soffit, lighting downwards. These are associated with concealed ground mounted floods located in landscaped areas at the bottom of the staircase towers and also in planted areas to the south side of the building.

The three-storey Entrance Canopy with its lower glazed porte-cochère extending outwards from the entrance, forms the main nighttime attraction, as during the day; leaving no one in any doubt as to where to arrive. The interior lighting seen through the glazed front allows light from the interior to spill outwards.

The car park and circulation areas are lit by 1 metre high bollards, containing a 70 watt SON T lamp, forming the main exterior lighting to the site. Low-level step lights are built into changes of level at ramps and steps. Additional lighting is added at certain landscape features, such as underwater lighting to the feature pool on the south side of the office, and ground mounted floods with mercury discharge lamps giving a change of colour to landscaping and trees.

The whole system is controlled from within the building, through the BEMS system, using photocell operation to turn the lighting on at dusk, and off at dawn.

Courtesy of Abbey Holford Rowe

Yorkshire Building Society, view towards the entrance hall (Architect Abbey Holford Rowe)

Lighting designer DSSR, Manchester

Courtesy of Abbey Holford Rowe

Yorkshire Building Society, entrance hall detail (Architect Abbey Holford Rowe)

Courtesy of Abbey Holford Rowe

Yorkshire Building Society, view of site showing pathway lighting (Architect Abbey Holford Rowe)

Baystreet Shopping Centre, Malta

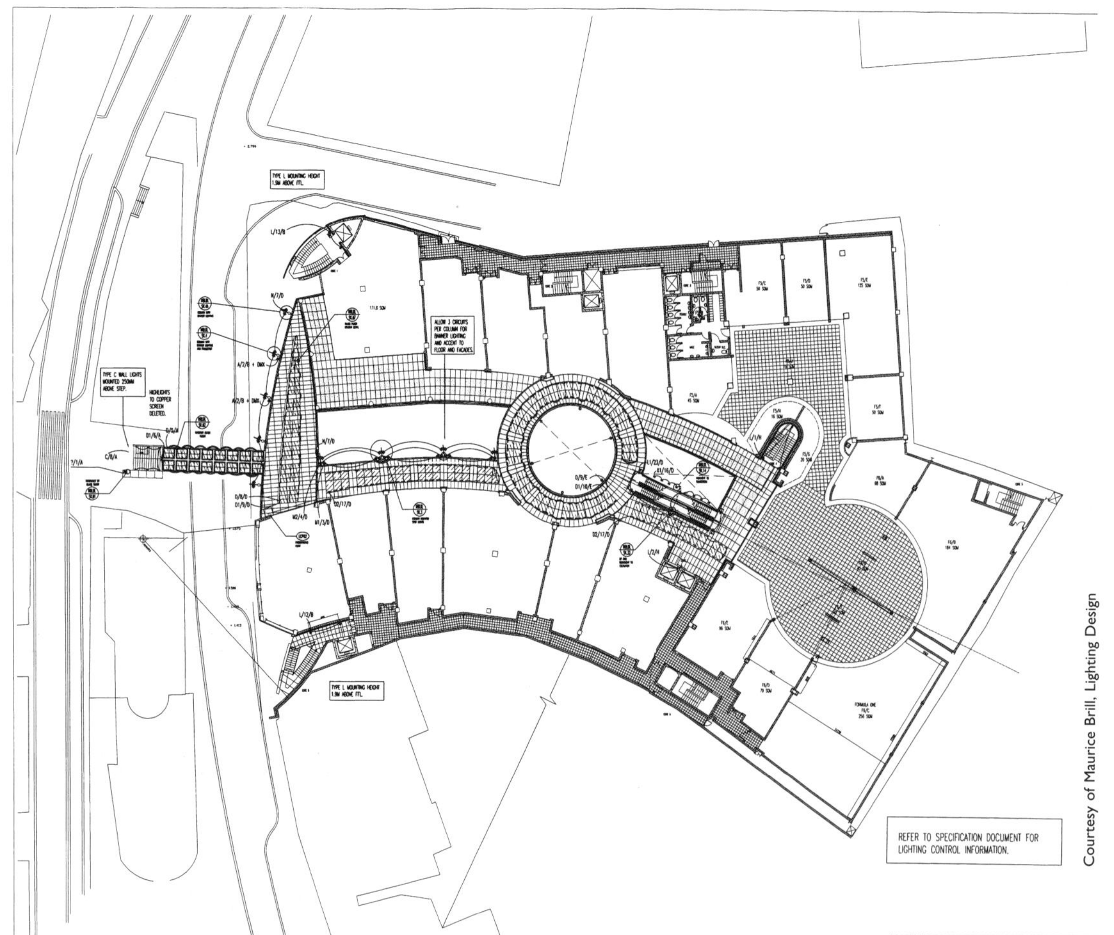

Courtesy of Maurice Brill, Lighting Design

Baystreet Shopping Centre, Malta, plan

Courtesy of Robert Honeywill

Baystreet Shopping Centre, Malta, the main entrance, night view (Architect BDG McColl)

Courtesy of Robert Honeywill

Baystreet Shopping Centre, Malta, open air access/lit walkways (Architect BDG McColl)

Lighting designer Maurice Brill, Lighting Design

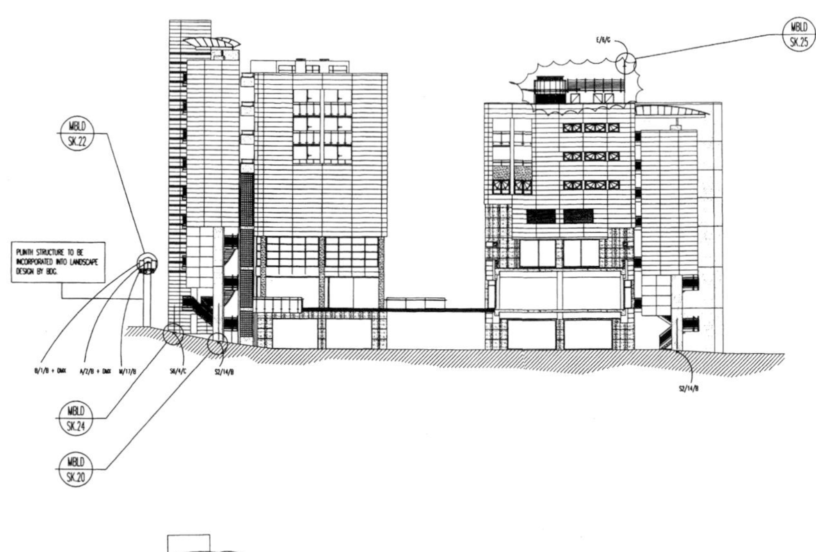

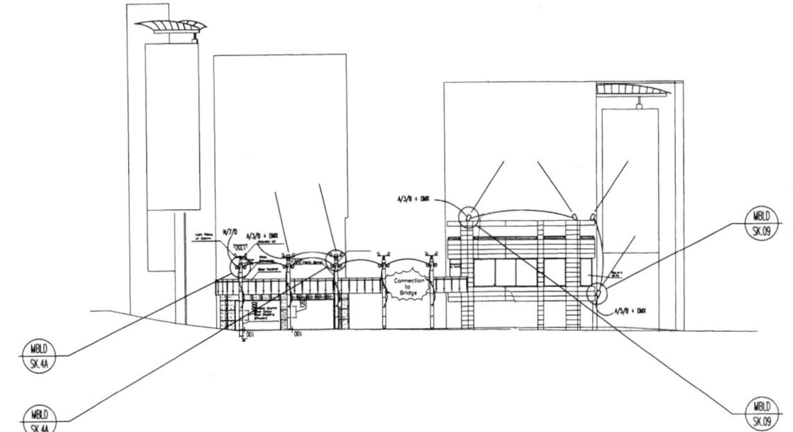

Baystreet Shopping Centre, Malta, section of the scheme

Courtesy of Maurice Brill, Lighting Design

Baystreet Shopping Centre, Malta, daylight view of rotunda lighting installation (Architect BDG McColl)

Courtesy of Robert Honeywill

In November 2000, Bay Street was opened in an up-and-coming suburb of Valetta on the main island of Malta. Described as a 'shopping and entertainment centre', the complex contains shops, restaurants and facilities for outdoor spectacles.

The climate of Malta being mediterranean, the problem for the architect was not so much to protect the public from cold and rain, as it was to protect them from the strong daylight, the heat and the sun. For this reason all the main circulation is on the outside, giving access to three levels of air-conditioned accommodation above. An important feature of the outside environment is the main atrium or rotunda used for occasional theatre, where exterior lighting is provided (see daylight photograph). In addition, the stone floor has blue and green LEDs set to achieve a changing pattern of concentric rings, introducing a theatrical spinning colour change on occasions.

The lighting design for the complex complements and emphasizes the main architectural features: designed to attract, lead and entertain the visitor. By subtle changes of mood, the atmosphere slowly changes from daytime shopping through the evening to late night dining and parties, in which the lighting plays a material part.

Front façade

The façade is washed by colour projectors, which fade through a series of colours to give a constantly changing appearance, and at weekends this is supplemented by 'Gobos' adding to the excitement.

Access bridges and staircases

Much use is made of blue and purple cold cathode tubes uplighting the frosted glass strips comprising parts of the walkways, and these give a soft light to the exterior circulation areas.

Escalators

The escalators which take the public from the second floor rotunda to an event terrace at roof level, are provided with surface mounted up and down lights, while the sunblinds which protect the escalators during the day are washed with amber light at night. Although the lighting can be described as theatrical, it uses low power consumption, and fittings which can be maintained easily to ensure that the running costs are not excessive.

Hotels and leisure

Tivoli Gardens, Copenhagen

The lighting of Tivoli Gardens has been referred to in the Introduction chapter as an example of a lit environment which has developed over many years, to the delight of the inhabitants of Copenhagen, and to the many visitors. The original garden was founded by Georg Carstensen in 1843, but the nighttime lighting has been developed over the last 50 years from the middle of the twentieth century, being regularly updated and improved, so that visitors will be conscious of a gradual change in the appearance of the different spaces over time.

The 8 hectare wooded site in the middle of the city, unlike Disneyland, does not rely on architectural superficiality for its appeal. All the various buildings, restaurants, concert hall and lakes fulfill the purpose for which they were designed, despite taking an eclectic view towards the different styles of architecture from which inspiration was drawn.

It is unnecessary to go into the technical details of the various methods of lighting. Suffice to say that the most up-to-date technology has been adopted to give the impression of a magical environment. Examples of this are the dragonflies on the lake which, due to the lighting, appear to hover; while in the background the three-masted frigate, *St. George 3*, lies at anchor in the lake, lit by subtle uplighting to ensure that it can be seen as a part of the total experience of the environment. Also in this illustration can be seen the small wooden bowl fountains, set among the planting mentioned earlier (see in Chapter 4, Fountains).

Tivoli has always been at the forefront of lighting innovation, and the hanging lights over another of the lakes are very much in the tradition of the designer Paul Henningsen whose interior lights designed for Louis Poulson many years ago have become classics.[2] Lighting and water play a major part, as can be seen in the tree lighting and the many fountains featured throughout the gardens; those in front of the Concert Hall being particularly effective. In all, the lit environment of Tivoli is a delight, and would have been worthy of the 'stately pleasure dome' decreed by Kubla Khan.

Courtesy of Louis Poulson

Tivoli Gardens, Copenhagen, view of Frigate *George 3*, with wooden bowl fountains

[2]Phillips, D. (2000) *Lighting Modern Buildings*, p. 71. Architectural Press.

Lighting designer Tivoli, Louis Poulson with Poul Henningsen

Courtesy of Louis Poulson

Tivoli Gardens, Copenhagen, dragonflies on the lake

Courtesy of Louis Poulson

Tivoli Gardens, Copenhagen, tree lighting, miniature sparkle lamps

Courtesy of Louis Poulson

Tivoli Gardens, Copenhagen, hanging lights

Approaches to the Millennium Dome, London

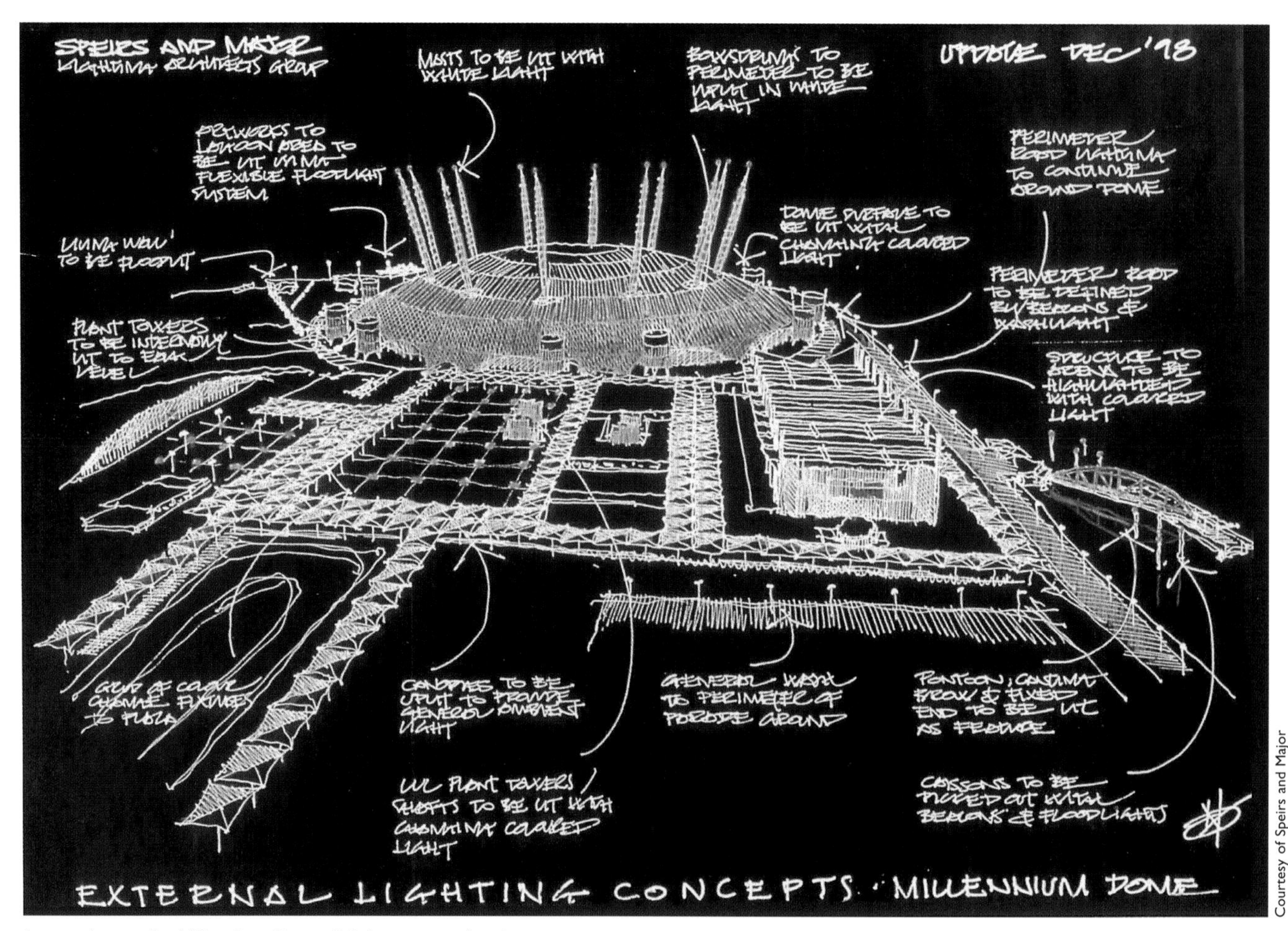

Courtesy of Speirs and Major

Approaches to the Millennium Dome, lighting concept for the area

DP Archive

The Millennium Dome, view of the Dome across the main square (Architect Richard Rogers Partners)

Lighting designer Speirs & Major

DP Archive

The Millennium Dome, daylight view of canopies (Architect Richard Rogers Partners)

DP Archive

The Millennium Dome, light fitting up to canopies (Architect Richard Rogers Partners)

DP Archive

The Millennium Dome, night view of canopies (Architect Richard Rogers Partners)

The main approach to the Millennium Dome, whether by bus or tube, is via the North Greenwich Underground station and the approach across the main square towards the entrance can be traced on the plan shown opposite.

The associated buildings, together with the dome, provide a significant amount of light to the area. The square itself has no ambient light but the main walkways are lit by special asymmetrical light fittings containing metal halide lamps. These fittings have sand-blasted diffusers which assist in improving the light distribution and light up to the translucent canopies which protect visitors from the weather, while providing a welcoming and attractive appearance to the area. On arrival, the visitor must negotiate the entrance kiosks, for payment and prepaid ticket collection. These kiosks are illuminated with white light when they are open, and blue light when they are closed.

The cylindrical plant towers have each of their internal levels up-lit using dimmed tungsten halogen fittings with red dichroic filters. These towers form a significant aspect of the surroundings of the dome and can be seen – as of course can the dome itself – from a considerable distance; the warmth of the light makes a contrast with the white surface of the dome.

The many food outlets in the area are provided with their own type of lighting suited to the style of restaurant, adding variety to the scene. This also applies to the fair-ground activities; the lighting and architecture of the area being robust enough to ensure that these popular entertainments are subservient to the main attraction, the dome itself.

Hilton Hotel, Tel Aviv, Israel

The lighting design for the pool terrace at the Hilton Hotel in Tel Aviv in Israel was coordinated with the landscape architect in charge of the redevelopment of the whole area – his design allowed for a previous rectangular chalet building to be replaced by curving terraced lawns and palm trees.

It was clear that the new design called for a change of approach by producing a lighting scheme that was both sympathetic to the landscape and at the same time modelled the terrace in a subtle and integrated manner.

The terrace provides views out to the Mediterranean, and it was clearly important that these views should be undisturbed by the lighting pattern. The design should be well controlled and avoid the easy solutions of floodlighting the terrace from the Hotel buildings, or achieving uniform light distribution by means of high level columns.

The layout of the terrace is indicated on the diagram, and shows that all the pathways are lit by louvred brick lights containing 18 watt CFLs recessed into the low-level side walls, together with fittings containing 10 watt capsule lamps lighting the steps.

The two pools are lit by underwater integrated lighting and each of the palm trees have two recessed buried uplighters (35 watt CDM-T lamps) associated with them.

The scheme is completed by a series of light sculptures consisting of five groups of special lighting bollards at varying heights (1.2 metres × 2.7 metres and 4 metres height) these provide pools of light at strategic positions, such as changes of level ensuring the safety of guests, whilst at the same time achieving a dramatic architectural feature. It is perhaps wrong to describe these sculptures as bollards, as they are a special design by DZ Licht and their appearance is more of a lit sculpture.

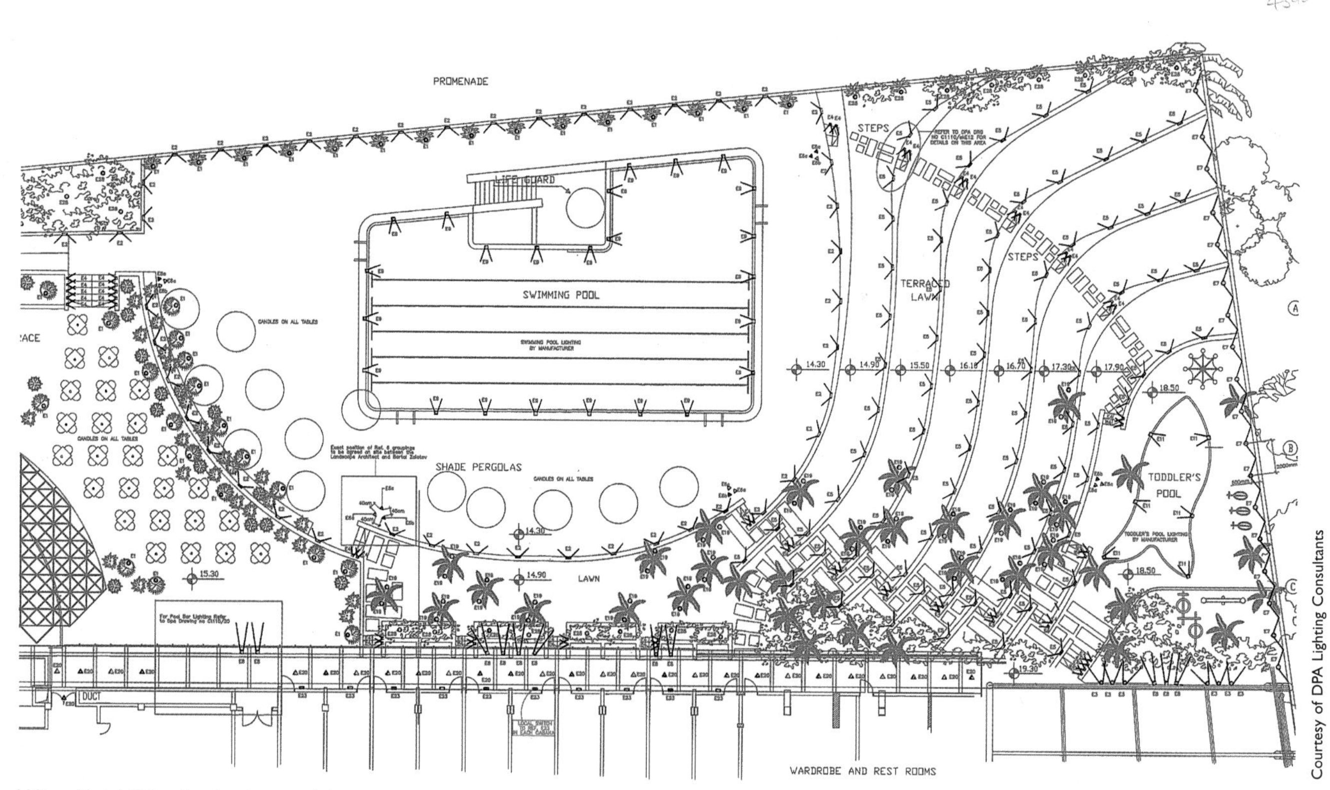

Hilton Hotel, Tel Aviv, plan layout of the area

Courtesy of DPA Lighting Consultants

Lighting designer DPA Lighting Consultants

Courtesy of DPA Lighting Consultants

Hilton Hotel, Tel Aviv, palm tree lighting with pool beyond (Architect Yacov Rechter; Landscape architect Uri Mueller)

Courtesy of DPA Lighting Consultants

Hilton Hotel, Tel Aviv, detail of the lit sculptures (Architect Yacov Rechter; Landscape architect Uri Mueller)

Courtesy of DPA Lighting Consultants

Hilton Hotel, Tel Aviv, view of the main pool (Architect Yacov Rechter; Landscape architect Uri Mueller)

Cutty Sark Gardens, London

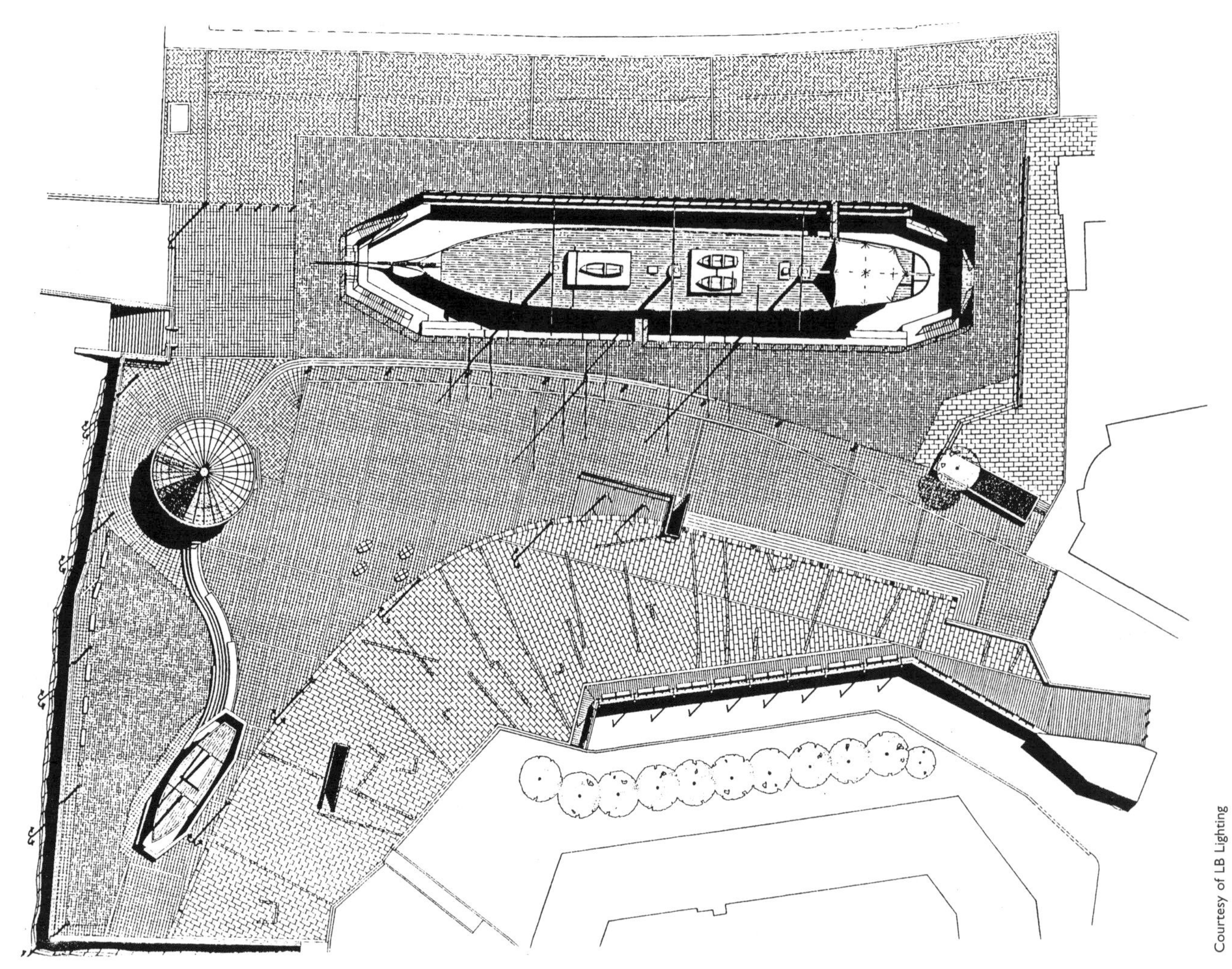

Courtesy of LB Lighting

Cutty Sark Gardens, plan layout of the gardens

Courtesy of Timpson Manley

Cutty Sark Gardens, general view of the gardens at night (Architect Timpson Manley Ltd)

Lighting designer LB Lighting with Fulcrum Consulting

Courtesy of Timpson Manley

Cutty Sark Gardens, night view of the seating (Architect Timpson Manley Ltd)

Courtesy of Timpson Manley

Cutty Sark Gardens, day view of the area (Architect Timpson Manley Ltd)

At the heart of the maritime quarter of Greenwich beside the Thames, the gardens were for many years a barren and windswept space, dominated by exposed concrete surfaces and a semi-underground car park.

Today the original clipper ship, the *Cutty Sark*, has been restored in a proper setting, and this together with Francis Chichester's *Gypsy Moth IV* are set into a welcoming public space, worthy of London's historic maritime quarter. Neither of the ships is lit at night, and the *Cutty Sark* particularly would benefit from this, as the *Victory* does in the Historic Dockyard in Portsmouth.

The project, a major face-lift, involved repaving the whole area with coarse textured stone and Douglas Fir timber decking and the broadwalk with new seating, signage and lighting.

An important element of the lighting scheme is the glass dome to the foot tunnel rotunda which is emphasized by green light to the dome. The western side of the gardens has been provided with a timber screen interpretation wall incorporating graphics and lighting. Timber benches relate to this new screen wall and below-seat lighting is added with illuminated banners above.

The lighting is completed by low-level bollards outlining the curved pattern of paving, with heritage street-lights giving additional light towards the river frontage.

The scheme was entered for and won a Civic Trust Award for 'An outstanding contribution to the quality and appearance of the environment'.

Hyatt Regency Hotel, Thessaloniki, Greece

The lighting design was developed in close collaboration with the landscape architects, and leads the eye down the slope of the hillside from the hotel restaurant towards the main pool and pool deck below.

The restaurant terrace is lit using MR16 50 watt up-and-down lights placed between the large arched window openings of the building. Since the terrace is used for nighttime dining, this provides sufficient ambient light associated with candlelight to the tables. The main pool is lit with sealed underwater lights employing the low voltage 12 volt PAR 56 300 watt lamp, located in the sides of the pool; these can be removed and serviced while the pool is full of water. In another area a rock wall is also lit by means of underwater lighting which, due to the turbulence of the water, conceals the fittings. The subtlety of the lighting scheme is indicated by the concealed lighting incorporated in the rock faces leading to the changing rooms from the pool deck, which otherwise might have appeared rather threatening.

The main landscape lighting consists of lighting to the palm trees, together with low-level accent light to rocks, and small landscape features. The trees are lit using buried uplights (500 watt PAR 56 lamps) and pathways use the small low voltage capsule lamp (12 volt 50 watt MR 16) concealed so as to avoid glare.

The night experience is designed to be both stimulating and romantic.

Courtesy of Robert Honeywill

Hyatt Regency, Thessaloniki, restaurant terrace (Architect Thymio Papayannis & Associates; Landscape Architects Derek Lovejoy Partners)

Lighting design Maurice Brill, Lighting Design

Courtesy of Robert Honeywill

Hyatt Regency, Thessaloniki, view towards the pool (Architect Thymio Papayannis & Associates; Landscape Architects Derek Lovejoy Partners)

Courtesy of Robert Honeywill

Hyatt Regency, Thessaloniki, tunnel to the changing rooms (Architect Thymio Papayannis & Associates; Landscape Architects Derek Lovejoy Partners)

Temple of Luxor, Egypt

Looking at the pictures of the nighttime lighting of the Temple of Amon at Luxor, one might be forgiven for thinking that 'I could have done that!' – because it looks so simple – 'a one golf club solution'. Not so! This is, in fact, a very sophisticated and thoughtful lighting scheme carried out with a sense of history.

The programme launched by President Mubarak to revitalize northern Egypt included the nighttime lighting of the entire Temple at Luxor and also its reconstruction – rebuilt almost entirely from the stones which had fallen in the past. Students of architecture will have seen pictures of this dilapidated temple in Sir Bannister Fletcher's book *A History of Architecture*[3] and from this one can get some idea of the extent of its reconstruction; it had taken over 100 years to build, and yet was reconstructed in far less.

It is difficult to imagine the size of the elements of the project; the dromos (pathway) leading from the Temple of Karnak up to the gateway pylons to Luxor used for ceremonial processions was 2.5 kilometres long, the Gateway itself marked by a 25 metre obelisk, whilst the entrance pylons extend horizontally for 56 metres, and rise 30 metres in height. The frontage to the pylons have bas-relief sculpture which illustrate military campaigns, and it was important that these be emphasized at night by the lighting design.

In the words of the lighting designer for the project:

> The lighting from below exhalts the majesty of the setting, pulling out of the dark elements that are representative of a mysticism which inspires respect and admiration even in our day and age.

In the days when the Temple was built some 1500 years before Christ, almost certainly the reaction would have been one of awe and fear rather than respect, representing as it did the power of the priesthood and their hold over life and death.

It might be said that the designer was making a virtue of necessity, in that it would be difficult to see a method of lighting other than by means of floodlights mounted at low-level lighting upwards, that would have blended with the architecture.

The tall columns lining the courtyard of Rameses are lit using the 150 watt metal halide lamp (warm, 3000K) with a closely controlled beam to avoid light spillage. The two colossal granite statues which flank each side of the entrance pylons are also lit by the metal halide lamp, with a combination of 400, 250 and 150 watt lamps illuminating the frontage and providing compensating light to the sides.

The sophistication of the lighting design lies in its ability, by means of a combination of fittings and light sources, to enable a visitor to appreciate the architecture, not as experienced in 1500 BC, but as a present-day tourist attraction.

Courtesy of I'Guzzini

Temple of Luxor, view of the temple by daylight

[3]Fletcher, Sir Bannister (1996) *A History of Architecture*, 20th edition. Architectural Press.

Lighting designer l'Guzzini

Courtesy of l'Guzzini

Temple of Luxor, entrance gates seen on arrival at night

Courtesy of l'Guzzini

Temple of Luxor, details of the columns at night

Marine

Brugge, the Canals

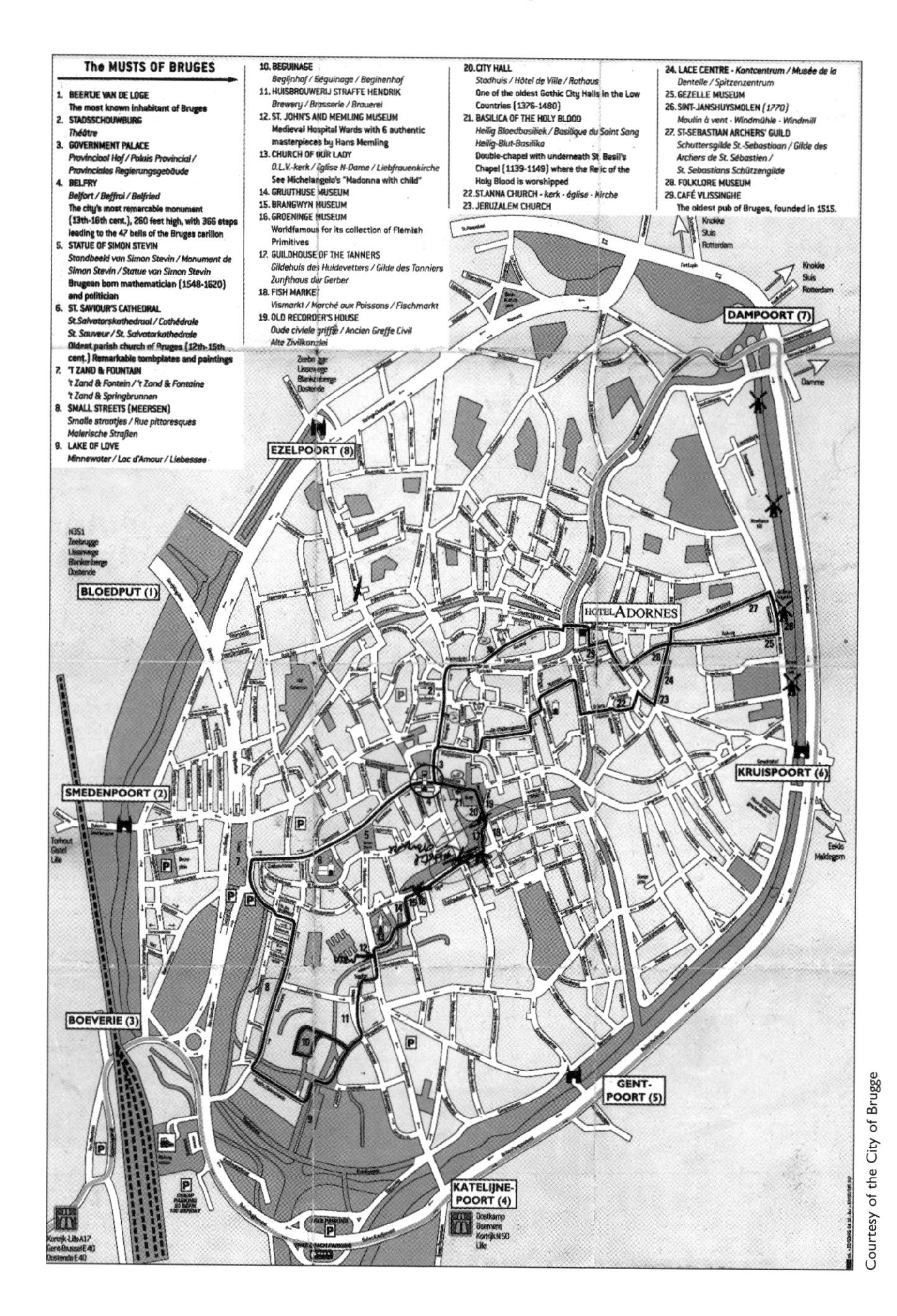

Courtesy of the City of Brugge

The Canals of Brugge, overall plan of the town with canals

The town of Brugge in Belguim is well known as a fine historic city with notable civic buildings. Churches, galleries, hotels and restaurants abound, but it is the canal system which both encircles and winds through the centre of the city which distinguishes it from towns in Britain – and it is the way in which the nighttime lighting of the canal environment has been achieved that is of importance.

The canals are for the most part lined with buildings, which assist in allowing lighting floods to be hidden from normal points of view, with floodlit views to be had from bridges, and adjoining roads. Where there is a unilateral view to a bridge or building, as on a stage, the lit appearance is most successful, but even when views of the equipment can be seen, considerable efforts have been made to reduce the effect of glare by means of placing the fittings in deep recesses in the opposite bank. Many innovative methods have been adopted to cut-off light from the views of the public, and this in the main succeeds.

There is a consistency of colour of the light sources used for the lit appearance, largely achieved by warm sources such as metal halide and HP Son. However, because of the individuality of the buildings along the canals there are a variety of appearances with cooler interior effects contrasting with the warmth of old brick and stone.

It would clearly be impossible to provide a comprehensive view of the canal environment so one or two views have been selected, as representative of the overall experience of the town and its canals. When visiting the town, for the greatest enjoyment choose the early spring or autumn to avoid the millions of tourists which will otherwise be there. As a tourist city it is a great success, not least because of the beauty of its nighttime appearance.

Lighting designer City of Brugge

DP Archive

The Canals of Brugge, view of bridge over canal

DP Archive

The Canals of Brugge, view of canal junction

DP Archive

The Canals of Brugge, view of canal with lit trees

HMS *Victory*, Portsmouth

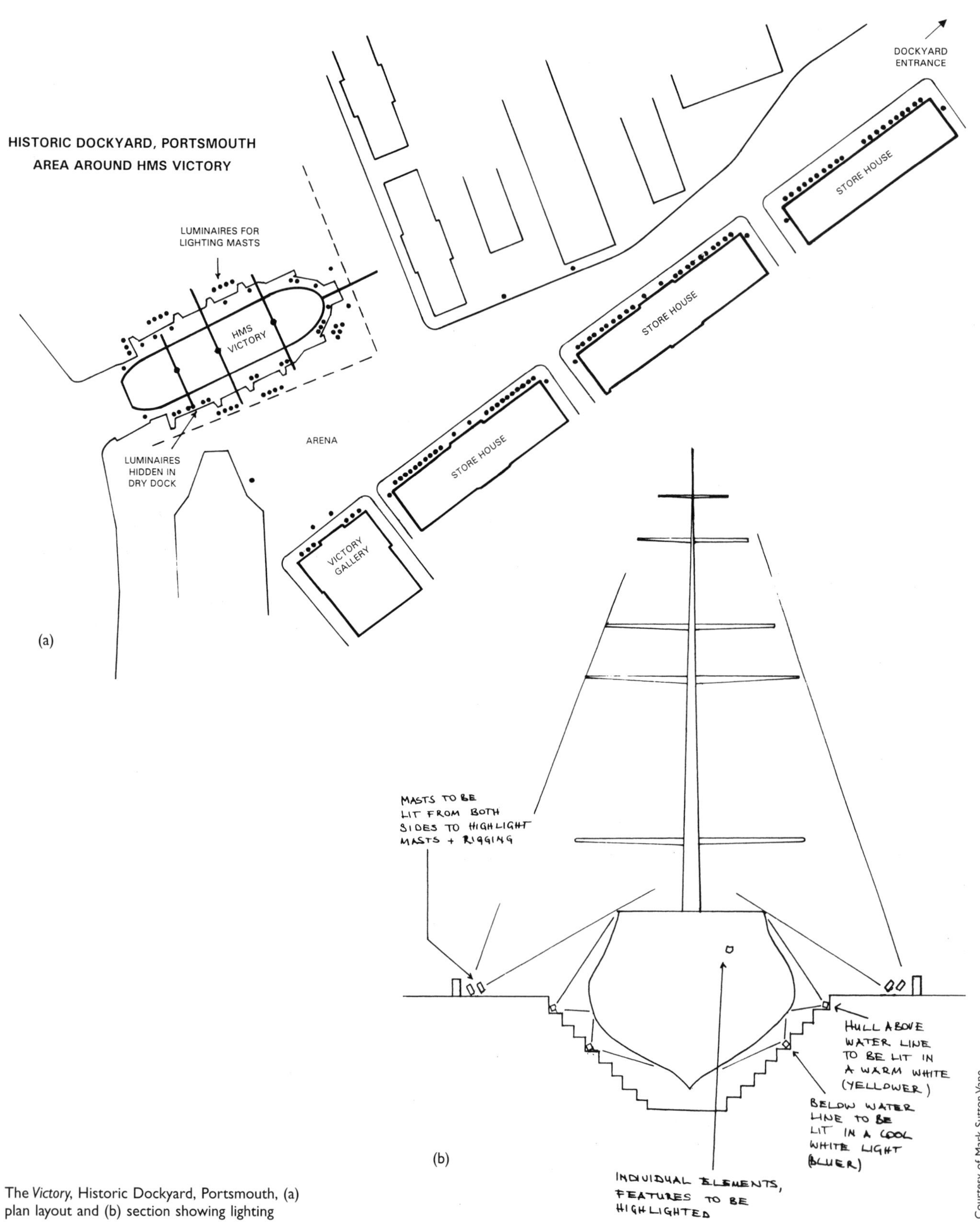

The *Victory*, Historic Dockyard, Portsmouth, (a) plan layout and (b) section showing lighting

Courtesy of Mark Sutton Vane

Lighting designer Sutton Vane Associates

Courtesy of Mark Sutton Vane

The *Victory*, Historical Dockyard, Portsmouth, floodlit (Landscape Architect Camlin Lonsdale; Client Portsmouth Naval Base Property Trust)

Courtesy of Mark Sutton Vane

The *Victory*, Historical Dockyard, Portsmouth, view of *Victory* with dockyard buildings (Landscape Architect Camlin Lonsdale; Client Portsmouth Naval Base Property Trust)

Courtesy of Mark Sutton Vane

The *Victory*, Historical Dockyard, Portsmouth, view of the bow of *Victory* (Landscape Architect Camlin Lonsdale; Client Portsmouth Naval Base Property Trust)

The lighting of Nelson's flagship in the Historic Dockyard in Portsmouth is very important. But it is also vital that the surrounding environment is lit in such a manner that it is safe and pleasant, without distracting the eye from the ship itself.

There is a wash of warm white light over the whole upper part of the hull, emphasizing the colourful quality of the wooden superstructure. Certain elements are picked out by individual spotlighting; these include the decorative figurehead and anchor. During the day the ship can be seen from a large part of the city, and this effect is reinforced at night by the nighttime lighting of the masts and rigging, carefully lit by narrow beam spots.

Cool white light is used to illuminate the lower parts of the hull, as a contrast to the warmer upper hull, and the metal rudder is given special emphasis by the use of green light. All light sources used are the metal halide CDM-T lamps for consistency of colour and good long-term economics. For conservation reasons no light fittings are attached to the ship itself, all fittings being located on the dock.

The lighting of the surrounding area is achieved by a combination of methods. The surrounding dockyard buildings are lit so that *Victory* is not the only lit object in view. The light reflected off the surrounding buildings provides the functional light to the open paved spaces, whilst various aspects of the buildings are lit to higher levels to emphasize the character of the architecture. In areas where the building lighting is insufficient, low modern design lamp-posts have been added to achieve a human scale. The whole area has been given an attractive modern landscape treatment.

Alençon, Banks of the River Sarthe

A lighting masterplan was carried out in 1992 for the improvement to the lighting of the centre of the town of Alençon in France. It was clear that the river would become a major factor in the appearance of the town at night, and landscape architects were appointed to prepare plans to carry forward proposals for the right bank. These were completed in 1994.

Working closely with the landscape architects, the lighting designer identified a number of important features along the length of the riverbank. The first of these was the point at which the river makes a big loop framed by two bridges – the views from the bridges contribute different aspects to the appearance of the town.

The bridges are lit from below, emphasizing the solid masonry of the arches, and reflected in the water.

The appearance at night has been achieved by the lighting design to enhance the improvements to the landscaping, and views towards the town, while creating attractive conditions for pedestrians by means of floodlit trees and illuminated pergola.

A special feature has also been made of the historic washing places – although no longer used for their original purpose, they are roofed over and act as a refuge from the weather. The lighting inside forms lit incidents along the way at night and is a key factor in the success of the proposals, as also is the use of the low-pressure sodium lamp to give a strange appearance to those entering.

The Riverside walk is lit by means of slender standards, in the form of luminous columns containing 36 watt fluorescent lamps with warm colour rendering (3000 K). These lead the eye along the path, to ensure the safety of the public. The whole appearance provides a riverside scene which greatly enhances the quality of the town.

Courtesy of Roger Narboni, Concepto

Banks of the River Sarthe, Alençon, view of the bridge at Alençon (Landscape architect Agence Latitude Nord; Client Ville d'Alençon)

Lighting designer Roger Narboni, Concepto

Courtesy of Roger Narboni, Concepto

Banks of the River Sarthe, Alençon, illuminated pergola (Landscape architect Agence Latitude Nord; Client Ville d'Alençon)

Courtesy of Roger Narboni, Concepto

Banks of the River Sarthe, Alençon, riverside washing place (Landscape architect Agence Latitude Nord; Client Ville d'Alençon)

The Green Corridor, the River Sevres, Niort, France

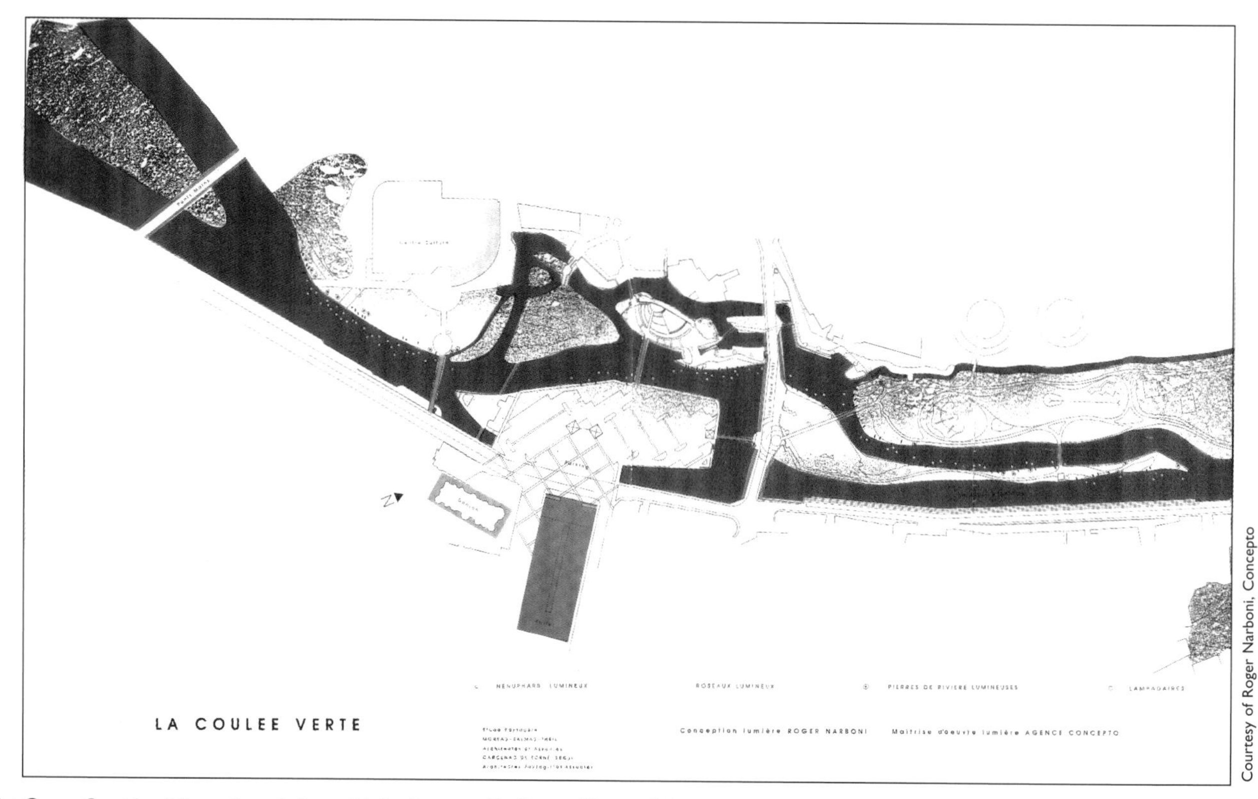

Courtesy of Roger Narboni, Concepto

The Green Corridor, Niort, view of river with lit elements (Architects Moreau Salmas and Theil; Landscape architects Jacques Segui and Carsenac de Torne; Client Ville de Niort)

Courtesy of Roger Narboni, Concepto

The Green Corridor, Niort, seen from the river bank (Architects Moreau Salmas and Theil; Landscape architects Jacques Segui and Carsenac de Torne; Client Ville de Niort)

Lighting designer Roger Narboni, Concepto

Courtesy of Roger Narboni, Concepto

The Green Corridor, Niort, view of the castle keep and covered market (Architects Moreau Salmas and Theil; Landscape architects Jacques Segui and Carsenac de Torne; Client Ville de Niort)

Courtesy of Roger Narboni, Concepto

The Green Corridor, Niort, cast iron bridge (Architects Moreau Salmas and Theil; Landscape architects Jacques Segui and Carsenac de Torne; Client Ville de Niort)

The town of Niort in the French Department of Deux Sevres, is one of the gateways to the coastal marshes of the Poitevin.

In 1988 the Mayor decided that the town had turned its back on the river, which had become a no-go area at night, and that the heritage of the river, its banks and associated vegetation was a treasure waiting to be discovered. Architects and Landscape Architects were commissioned to redevelop the banks of the river and to create a 'Green Corridor' as far as the town centre. At the same time the Mayor entrusted the lighting designer Roger Narboni with a study to illuminate this whole stretch of the Sevres, a length of 1.5 kilometers.

The Green Corridor is notable for its rich waterside vegetation with many islets and peninsulars, together with a sixteenth century Castle Keep, a nineteenth century covered market, and stone and wooden bridges; in all an ideal environment for a nighttime riverside walk.

A sophisticated comprehensive lighting scheme was put forward and completed in 1990, consisting of the following elements:

1. Floodlighting to the Castle Keep associated with the Covered Market, lit from inside with blue light
2. Lighting to the stone bridges and metal foot bridges
3. Submerged projectors giving the impression of luminous water lillies on the surface of the river
4. Tree uplighting along the banks
5. Luminous reeds, emitting narrow vertical lines of light, reminiscent of bullrushes
6. River stones placed along the riverside, concealing sources uplighting overhanging planting.

Many of these ideas have been incorporated into the designer's scheme, the submerged waterlilies being particularly effective. Lit in this way, the originality of the setting of the green corridor has transformed a no-go area of Niort into a friendly atmosphere in which the nighttime scene invites the visitor, as well as the locals, to take a relaxing stroll.

Harbour redevelopment, Bristol

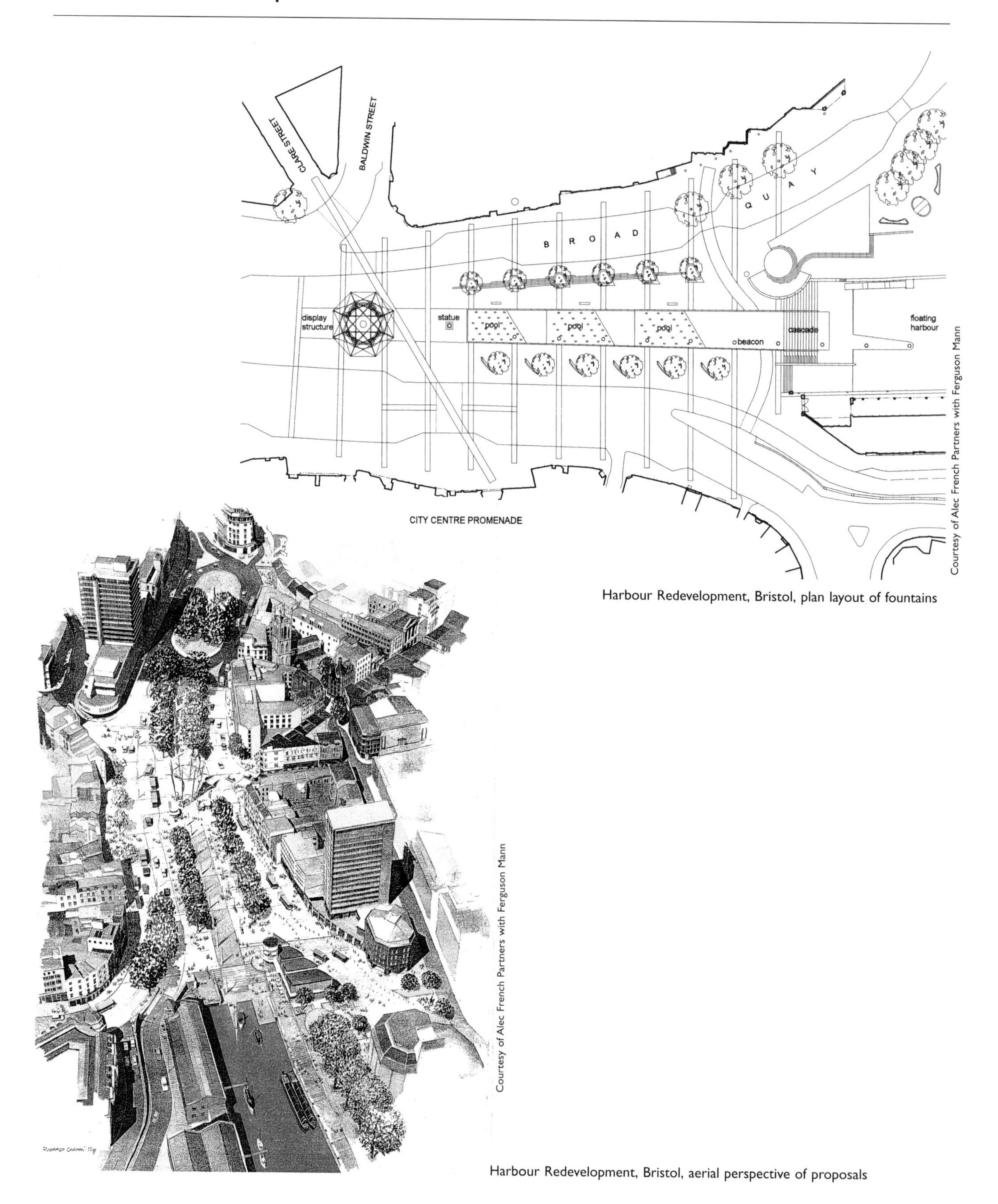

Courtesy of Alec French Partners with Ferguson Mann

Harbour Redevelopment, Bristol, plan layout of fountains

Courtesy of Alec French Partners with Ferguson Mann

Harbour Redevelopment, Bristol, aerial perspective of proposals

Lighting designer Alec French Partners with Ferguson Mann

DP Archive

Harbour Redevelopment, Bristol, general view of fountains, with lighting columns (Architect David Mellor of Alec French Partners, with Ferguson Mann; Engineer Ove Arup and Partners)

DP Archive

Harbour Redevelopment, Bristol, row of shops lining harbour (Architect David Mellor of Alec French Partners, with Ferguson Mann; Engineer Ove Arup and Partners)

DP Archive

Harbour Redevelopment, Bristol, new pedestrian bridge (Architect David Mellor of Alec French Partners, with Ferguson Mann; Engineer Ove Arup and Partners)

An area which was originally devoted to traffic has been regained for the city by an imaginative scheme linking the town centre to the floating harbour. This achieves the mixed development of water, buildings, people and activity reminiscent of the exciting port of Bristol in the nineteenth century.

The most striking feature is the linear fountain which originates from the centre of town down towards the floating harbour, with water jets lit by fibre optics providing a variety of cascades and colour.

Lining the fountains are a series of illuminated stainless steel columns or 'Millennium Beacons' – the design was entered, and won, in a competition by the artist Martin Richman. The simple shape of the columns seen during the day belies the hi-tech nature of the lighting design. They have four different lighting modes, the changing pattern of light giving impressions of solidity or transparency, whilst variations of colour generate an ever-changing light show.

The linear fountains flow towards the floating harbour, ending with a series of steps associated with a lit cascade.

Old warehouse buildings have been converted into restaurants and shops, from which access can be gained to Anchor and Millennium Squares (see the City Centre section above). Beyond this, a new pedestrian bridge engineered by Ove Arup Associates crosses the harbour, with a dramatic lifting section allowing tall ships through.

This is yet another example where lighting combined with imaginative planning and structure has converted a run-down area of what had once been a major seaport, into a lit environment which the public can enjoy.

DP Archive

Harbour Redevelopment, Bristol, cascade at steps to floating harbour (Architect David Mellor of Alec French Partners, with Ferguson Mann; Engineer Ove Arup and Partners)

Industrial

Zollverein Kokerie, Essen

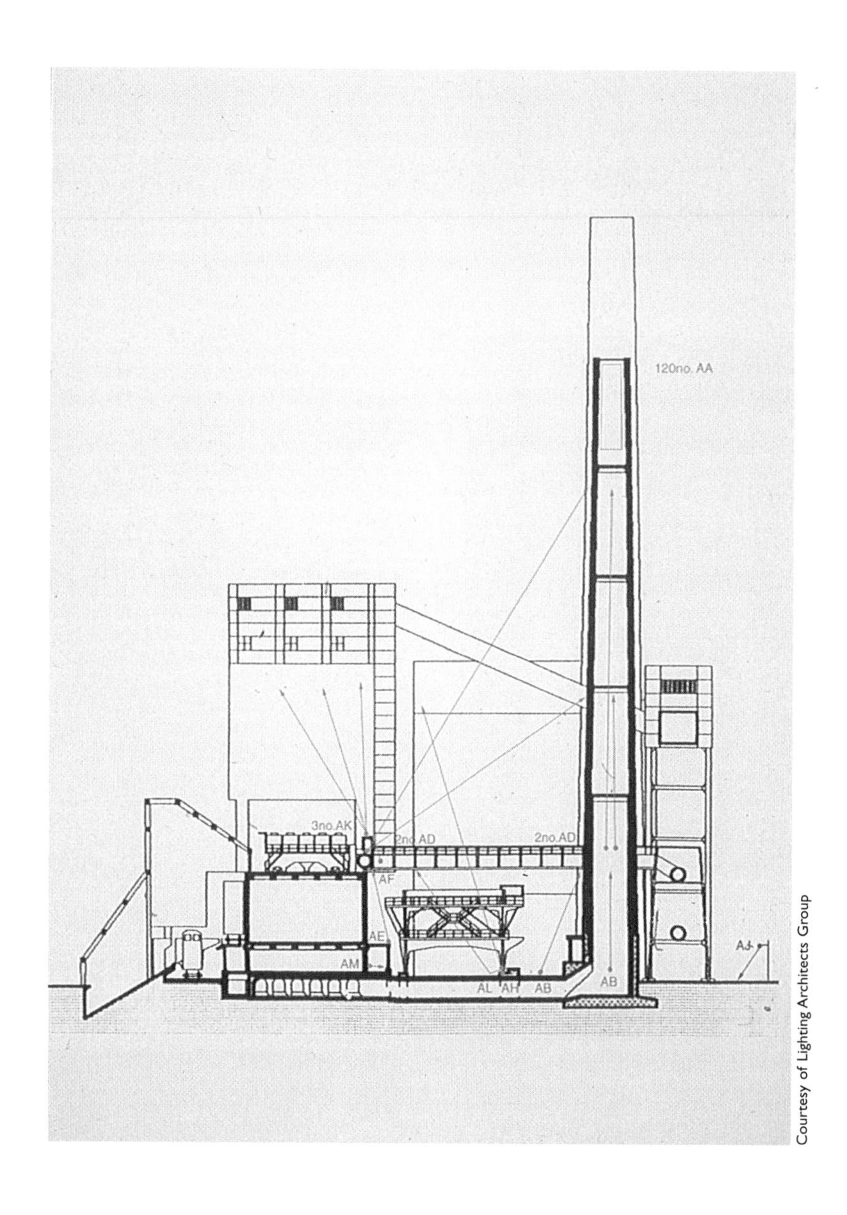

Courtesy of Lighting Architects Group

The transformation of a defunct coking plant into a monumental landmark for the Interbau Austellung (IBA) near Essen in Germany is a celebration of its industrial past; the ¾ mile long plant becomes a monumental industrial sculpture.

To create a stunning impact, visible 20 kilometres away, the installation is limited to two colours (red and blue) of light derived from metal halide sources, with colour filters. From a distance, the high chimneys, two at 75 metres and two at 60 metres, lit by clusters of LEDs, glow with light. But perhaps the most impressive aspect of the installation is the mirrored reflecting pool, which runs the entire length of the old works.

This pool is formed of aluminium sheeting tanked with black bitumen, and reflects the red façade of the building, doubling the visual impression; yet another example of the powerful effect of light with water.

The building was only used as an industrial exhibition for a year (the original structure had once stockpiled large quantities of coke). The exhibition gallery gave visitors an insight into the original use of the building, while views out onto the reflection pool added to the visual excitement of the scene. The lighting scheme remains as a lasting monument to the life of the building and the industry of the Ruhr.

Zollverein Kokerie, Essen, section through the complex

Lighting designer Lighting Architects Group, Speirs & Major; Jonathan Speirs Associates

Courtesy of Lighting Architects Group and Werner Hannafel

Zollverein Kokerie, Essen, general view of the installation with reflection pool

Courtesy of Lighting Architects Group and Werner Hannafel

Zollverein Kokerie, Essen, view of the chimneys coloured red

Residential

Eldridge House, Bristol

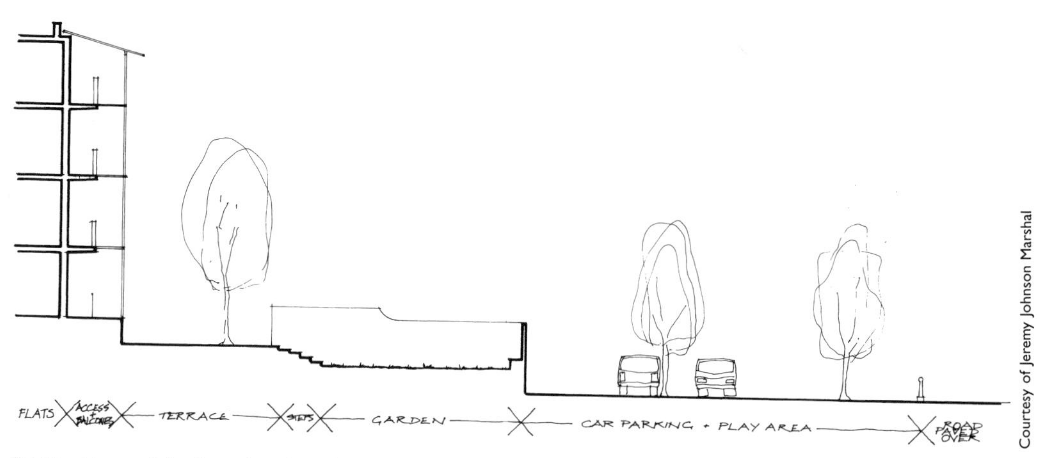

Courtesy of Jeremy Johnson Marshal

Eldridge House, Bristol, section through site

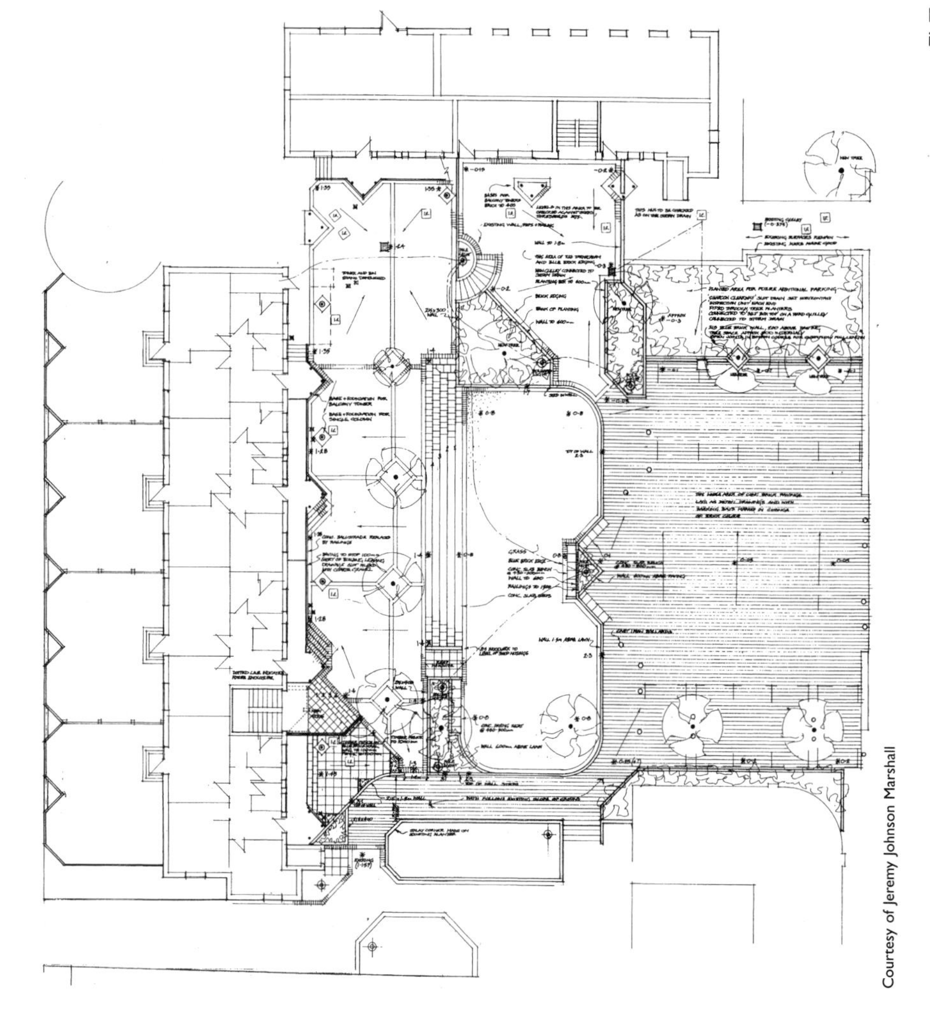

Eldridge House, Bristol, plan of site showing improvements

Courtesy of Jeremy Johnson Marshall

Lighting designer Jeremy Johnson-Marshall

Courtesy of Andrew Dykes Associates

Eldridge House, Bristol, view of flats before conversion (Architect Jeremy Johnson Marshall; Client Bristol City Housing Department)

DP Archive

Eldridge House, Bristol, completed scheme in 2001 (Architect Jeremy Johnson Marshall; Client Bristol City Housing Department)

DP Archive

Eldridge House, Bristol, view of flats from lower landscaped level (Architect Jeremy Johnson Marshall; Client Bristol City Housing Department)

These blocks of 1940s flats in a run down estate were due for environmental improvements to security and amenity; the works were carried out in 1990.

The major improvements were changes to the staircase access and balconies to achieve a degree of privacy and security to each tenant's flat, together with a wholesale redevelopment of the site. This achieved a separation between the people and the car, with the pedestrian layout of the estate changed to eliminate through-routes – previously a magnet to thieves – to create defensible spaces for the tenants.

It can be seen from the section that the cars are now housed at a lower level well hidden from view of the flats by the landscaping, by means of raised grassed terraces, trees and other planting. In the 10 years since the work was carried out the planting has matured to produce a pleasing environment, much used by tenants and their children.

Improvements to the lighting were an important part of the redevelopment with all the flats being given their own identity with a light at each front door. The exterior spaces were provided with a modest but adequate scheme of exterior lighting, employing four small-scale pole lights, with cut-off lanterns. Each lantern contains two 24-watt CFC lamps. In addition, a number of low-level bollard lights were provided to enliven the planters dividing up the space.

While the main expense was the improvements derived from the changes to the architecture and the landscaping, the relatively small cost of lighting contributed disproportionately to achieving both a functional and pleasant environment.

CONCLUSION

The 'Lit evironment' is the culmination of the design process which must be undertaken to achieve a nightscape which is both delightful and functional.

Architects should have a major part to play in the design process, since they, together with the other main professionals (city planners, landscape architects and lighting designers) must have the vision required to see the problem as a whole, in order to achieve the balance between technical solutions and visual fulfilment.

Earlier chapters have dealt with the individual components: the Buildings (Chapter 2), the Spaces between (Chapter 3) and the Incidents (Chapter 4), which form the lighting opportunities, followed by the tactical questions of lighting engineering ... how much and how to provide it ... but in this final chapter, the Lit Environment has been discussed in terms of a number of nightscapes chosen to illustrate their visual fulfilment, and the variety of solutions available.

The design process

Since the design process cannot be said to start without it, the assumption is that a decision has been made that the area concerned will be lit at night, that there will be a planned nightscape. This decision will have been made by the town authorities and the planning officer, taking account of the situation on the ground in existing spaces, or the proposed plans where the spaces are as yet unbuilt.

In existing historic spaces, an agreed visual master plan may have been drawn up, to which the lighting proposals will be subject. However, where there is no such plan, or where it is new construction, it is important that a survey is carried out to establish all the factors which will contribute to the final design.

These factors would include the nature of the site, and its immediate surroundings, and the functional needs of those who will use it. The needs of motor traffic as well as the needs of the pedestrian and cyclist must be investigated and flow patterns developed.

How will the space be used? If it is existing it is likely that a renewed nightscape will increase its popularity and therefore the numbers of people to be accommodated. If as yet unbuilt, a generous allowance should be made to cater for future numbers, and activities feeding upon success.

Such a survey should lead to a performance specification, to which the architect and his team should adhere; the difference being that such a specification should concern itself not only with the functional aspects of safety and security, but with the visual aspects of the overall appearance and ambience of the site.

Buildings will often be the defining edges of a space, and for this reason, the way in which they are experienced – how the façades are modelled, whether lit externally by floodlighting or from their own interior lighting – can have the greatest impact on the space.

Likewise, the manner of the lighting to the space needs to achieve a human scale. This will not be achieved by treating the space like a prison yard as so many spaces in our towns have been treated in the past, with high levels of uniform light from tall lighting columns, on the assumption that we are all a nation of vandals. Such an attitude ensures that if we are, we will remain so, but gives little opportunity for hope and improvement.

DP Archive

Venice by moonlight

It cannot be over-emphasized that 'a little light goes a long way at night' and that the pools of light derived from carefully planned lit incidents such as planting, fountains, trees or seating, can often be all that is needed.

It is important to remember that on completion the final space will be for the enjoyment of the public who will use it; the lighting will need to assist in human activity whether this is in some form of sport, shopping, or simply walking the space or sitting out. In the final analysis the lit environment must be somewhere that people wish to visit and return to time and again, it must be neither gloomy or threatening nor uniformly bright – in short, it requires a balance.

It is hoped that the lit environments listed will have demonstrated the quality which can be achieved, where a human view of the visual environment has been considered; where the lit opportunities are embraced, and the advantages of nature in planting and vegetation, or the delights of water have been seized, and 'where every prospect pleases'.

Bibliography

BOOKS

Bannister, Sir Fletcher, *A History of Architecture,* 20th edition, Architectural Press, 1996
Buchanan, C. C., *Traffic in Towns,* The Buchanan Report, 1963
Gardner, Carl and Hannaford, Barry, *Lighting Design,* Design Council, 1993
Phillips, Derek, *Lighting in Architectural Design*, McGraw-Hill, 1964
Phillips, Derek, *Lighting Historic Buildings,* Architectural Press and McGraw-Hill, 1997
Phillips, Derek, *Lighting Modern Buildings*, Architectural Press, 2000
Rasmussen, Steen Eiler, *Experiencing Architecture,* MIT Press, 1959
Schlor, Joachim Schlor, *Nights in the Big City,* Reaktion Books, 1998
Tregenza, Peter and Loe, David, *The Design of Lighting*, E & FN Spon, 1998
Wolfgang, Schivelbusch, *Disenchanted Night,* University of California Press, 1988

INSTITUTION PUBLICATIONS

CIBSE *Code for Interior Lighting*, 1994 (Update shortly)
CIBSE *The Outdoor Environment*, Lighting Guide LG6, 1992
CIBSE *Sports*, Lighting Guide LG4, 1990
CIBSE/ILE *A Guide to Good Urban Lighting*, 1995
CIBSE The Society of Light and Lighting (SLL) *City Lights,* The Waldram Lecture, 1990
European Commission, Directorate-General for Energy (Thermie) *Energy Efficient Lighting in Buildings: Offices/Industrial/Schools*
ILE *Lighting and Crime*, 1999
ILE National Lighting Seminar. *Developing Lighting Strategies – Essential 21st Century Planning*, 1999
ILE Guidance Notes. *The Reduction of Light Pollution* – Revised
LIF (Lighting Industry Federation) *Lamp Guide*, 1998
RIBA *Floodlighting Buildings (150 Years)*, 1983
Royal Fine Art Commission *Lighten Our Darkness*, 1993

PAPERS

Loe, David and Rowlands, E. S., *The Art and Science of Lighting: a strategy for Lighting Design,* LR & T/CIBSE 28, No 4, 1996

Phillips, Derek, *Space Time and Light in Architecture,* Presidential Address to the IES 1975 Trans Illum.Eng. Soc. 7, No 1, 1975
Phillips, Derek, *Architecture – Day and Night,* CIBSE National Lighting Conference, 1992

JOURNALS

Light & Lighting, magazine of the CIBSE/SLL
International Lighting, Review 50th Anniversary Special Edition. ILR/992
The International Lighting Review, published by Philips Eindhoven
The Lighting Journal, published by Institute of Lighting Engineers (ILE)
Light, published by ETP
Lighting Equipment News, published by EMAP
NYT, published by Louis Poulson Lighting A/S
Professional Lighting Design, published by European Lighting Designers' Association (ELDA)

Glossary

A simplified explanation of references used in the text divided into the following seven headings. Use the Index for page references. This is similar to the Glossary used in *Lighting Modern Buildings*,[1] which will enable the two books to be used together.

1. SEEING/PERCEPTION
2. LIGHT SOURCES/DAYLIGHT
3. LIGHT SOURCES/OTHER THAN DAYLIGHT (artificial)
4. LIGHTING TERMINOLOGY
5. LIGHTING METHODS
6. ENERGY AND CONTROLS
7. ARCHITECTURE

1. SEEING/PERCEPTION

Adaptation The human eye can adapt to widely differing levels of light, but not at the same time. When entering a darkened space from a brightly lit space, the eye needs time to adapt, to the general lighting conditions; this is known as 'adaptation'.

Clarity Clearness, unambiguous.

Contrast The visual difference between the colour or brightness of two surfaces when seen together. Too high a contrast can be the cause of glare.

Modelling The three-dimensional appearance of an object surface or space as influenced by light. Good modelling aids perception.

Perception Receiving impressions of one's environment primarily by means of vision, but also one's other senses; providing a totality of experience.

Quality, a degree of excellence The 'quality' of a lighting design derives from a series of different elements, the most important of which is 'unity', but which also includes aspects such as modelling, variety, colour and clarity.

Unity The quality or impression of being a single entity or whole; this can be applied equally to a small or large complex, the word 'holism' is often used in its place.

---

[1] Phillips, D. (2000) *Lighting Modern Buildings*, Architectural Press.

Variety The quality of change over time in brightness, contrast and appearance of a space, or series of spaces.

Virtual image An image of a subject or lit space formed in a computer, which can be used to provide a visual impression of the lighting design in order to explain a proposal.

Visual acuity A measure of the eye's ability to discern detail.

Visual task/task light The visual element of doing a job of work, and any local or concentrated light fitting placed to improve visibility.

2. LIGHT SOURCES/ DAYLIGHT

Bilateral daylight Daylight from both sides of a building.

Daylight The light received from the sun and the sky, which varies throughout the day, as modified by the seasons and the weather.

Daylight factor (DF) The ratio of the light received at a point within a building, expressed as a percentage of that available externally. Since daylight varies continually the amount of light from a given DF is not a finite figure, but gives a good indication of the level of daylight available.

Daylight linking Controls which vary the level of artificial light inside a building, relating this to the available daylight.

No sky-line The demarkation line within a building where, due to external obstruction and window configuration, no view of the sky is visible.

Obstruction/view The diminution of available light and view by other buildings at a distance. View is an important environmental aspect of daylighting, which may be impaired by obstruction, but can sometimes be overcome by attention to orientation.

Orientation The manner in which a building is related to the points of the compass. In the northern hemisphere care must be taken with southern exposure.

Shading/briese soleil The means adopted to prevent the deleterious effects of solar gain from southern exposures; these may be external structural louvres attached to the face of the building or forms of helioscreen blind.

Sky glare The unacceptable contrast between the view of the sky outside, and the interior surfaces.

Skylight The light received from the whole vault of the sky as modified by the weather and time of day, ignoring sunlight.

Solar gain Heat derived from the sun; whilst generally therapeutic, it may require control by forms of blind, louvre or solar glass.

Solar glass Glass designed to reflect a percentage of direct heat (infra red) from the sun.

Sunlight The light received directly from the sun, as opposed to that derived from the sky.

Sunpath The sun's orbit. As the earth travels around the sun, variations occur both throughout the day and the seasons; these changes in position are known as the 'Sunpath'. This can be accurately predicted.

Window 'Wind-eyes' take many forms, to provide daylight to an interior.

3. LIGHT SOURCES OTHER THAN DAYLIGHT/ARTIFICIAL

Arc light The first form of electric light derived by passing an electric current between two electrodes. Developed by Sir Humphery Davy in 1809.

Candles Candles are made by moulding wax or other flammable material around a wick, which sustains a flame to give light. Modern candles are clean, do not 'gutter', and provide light of a particular quality suitable for social occasions. There have been many light sources which attempt to imitate the quality of 'candle light'; most fail completely, while one or two later versions achieve some success.

Electric light The development by Edison and Swan of the 'incandescent' lamp in the nineteenth century and the arc lamp, providing the foundation of all modern forms of light derived from electricity.

Electric light sources These lamps are described in detail in Chapter 5, and are listed here.

Incandescent sources
tungsten filament
tungsten halogen
low voltage tungsten halogen

Discharge sources
cold cathode (fluorescent)
mercury fluorescent (high and low pressure)
low pressure sodium
high pressure sodium
high pressure mercury
metal halide (inc. ceramic arc)

Fluorescent lamps
halophosphor – tubular fluorescent triphosphor
compact fluorescent
induction lamps.

Fibre optics (remote source) At it simplest, it is the transfer from a light source placed in one position to light emitted in another, by means of glass fibre or polymer strands.

Fluorescent phosphors The internal coatings on surfaces of mercury discharge lamps which produce 'visible' light when excited by the ultraviolet rays emitted by the discharge. The phosphors determine the colour of the visible light.

Gaslight The light derived from burning coal gas, developed in the late eighteenth century, and widely used during the nineteenth century both for domestic and industrial use.

Oil lamps These together with firelight are the earliest forms of artificial light source, the oil being derived from animals, birds or fish. Hollowed out stone dishes and later clay pots were used with some form of wick. Oil lamps survived until the nineteenth century with the development of the 'Argand' lamp.

4. LIGHTING TERMINOLOGY

Angle of separation The angle between the line of sight and the light fitting. The smaller this is, the more likely the light will be glaring.

Brightness The subjective appearance of a lit surface; dependent upon the luminance of the surface and a person's adaptation.

Bulk lamp replacement The replacement 'en masse' of the lamps in a lighting system when it is calculated that a percentage of the lamps will fail, and that the light output of the system will fall below the design level.

Colour We accept that we only see true colour under daylight, despite the fact that this varies considerably throughout the day. All artificial sources distort colour in one way or another.

Colour rendering A comparison between the colour appearance of a surface under natural light and that from an artificial source.

Efficiency/efficacy The ratio of the light output from the lamp, to energy consumed in lumens/watt.

Flicker The rapid variation in light from discharge sources due to the 50 Hz mains supply, which can cause unpleasant sensations. With the development of high frequency gear the problem is overcome.

Glare/reflected glare The most important 'negative' aspect of quality. There are two types of glare, 'discomfort and disability.' Both types are the result of too great a contrast. Glare may result from both daylighting or artificial lighting, either directly or by reflection and must be avoided at the design stage.

Illumination level The amount of light falling on a surface expressed in engineering terms as lumens per square metre (or Lux) and known as 'illuminance'.

Intensity Refers to the power of a light source to emit light in a given direction.

Light fitting/luminaire The housing for the light source which is used to distribute the light. While the technical word is 'luminaire,' the more descriptive 'light fitting' is still widely used. The 'housing' provides the support, electrical connection and suitable optical control.

Luminance Light emitted or reflected from a surface in a particular direction; the result of the illumination level and the reflectance.

Lux The measure of 'illumination level' (illuminance) in Lumen/sq.m. The Foot Candle is used in the USA, meaning 1 lumen per square foot or 10.76 Lux.

Maintenance factor The factor applied to the initial illumination level, to take account of dirt accumulation and fall off in light output from the lamp, when calculating the level of useful light.

Reflectance The ratio of light reflected from a surface to the light falling upon it; as affected by the lightness or darkness of the surface.

Reflection factor The ratio of the light reflected from a surface, to the light falling upon it. The surface, whether shiny or matt, will affect the nature of the reflected light.

Scalloping The effect gained from placing a row of light fittings too close to a wall. Where intended this effect can be pleasing, but more generally it becomes an unwanted intrusion on the space.

Sparkle A word which may be applied to rapid changes to light over time, most readily applied to the flicker of candlelight or firelight; sparkle may be applied to reflected or refracted light from small facets, such as those of a glass chandelier.

5. LIGHTING METHODS

Ceiling/wall mounted The method by which light fittings are supported directly from the ceiling or wall.

Concealed lighting Concealed in the ceiling or wall configuration, to provide light on to adjacent surfaces.

Decorative lighting That which is designed to be seen and enjoyed for its own sake, such as a crystal chandelier. Alternatively it may be light directed on to objects to achieve a decorative purpose.

Downlight Light fittings giving their main light downwards; these are generally recessed and include both wide beam and narrow angles.

Emergency lighting The lighting system designed to operate in the event of power failure to facilitate the evacuation of a building, or continuation of essential services. Various methods adopted to ensure a suitable source of power.

Floodlighting Generally refers to the exterior lighting of a building at night, by means of lights with controlled beams placed at a distance.

Functional lighting Lighting which is planned to provide light to satisfy the practical needs of a space.

General Diffusing light fittings giving all round light.

Indirect Lighting provided 'indirectly' reflected from ceiling or wall.

Local light/task light A light fitting designed to light a specific task, generally at individual control.

Louvres/baffles A means of shielding the light from a fitting or from daylight, to eliminate glare. They can be fixed or moveable.

Portable light fittings Such as table and floor standards designed to provide local light. 'Portable' uplights a useful addition.

Raising and lowering gear The apparatus applied to heavy 'light fittings' in tall spaces, to allow them to be lowered for lamp change and maintenance.

Spotlight Light fittings designed to throw light in beams of varying width and intensity.

Suspended The pendant method of 'hanging' a light fitting from the ceiling or roof.

Torchere Originally a decorative free standing 'candle holder'; a term sometimes applied to modern wall brackets.

Track mounted light fittings Both supported and energized, from the numerous track systems available; giving flexibility.

Uplight Light fittings directing their light up to the ceiling providing indirect light.

Wall washing The means of lighting by which a wall is designed to be lit evenly; several methods can be adopted to achieve this, some more successful than others.

6. ENERGY AND CONTROLS

BEMS Building Energy Management Systems. A means of computer control of lighting systems within a building.

Control gear Discharge sources require 'control gear' comprising amongst others: starters igniters transformers, capacitors, ballasts and chokes to operate. Incandescent lamps require no gear, giving low initial cost and making dimming simple.

Digital multiplex controller A sophisticated electronic controller used to monitor and vary circuits in a lighting system, such as might be used in a theatre.

Dimming Dimming controls are exactly what the name implies, a device by which the intensity of a light source can be reduced. All filament sources, both mains and low voltage can be controlled by simple dimmers.

Intelligent luminaires Light fittings with inbuilt sensors programmed to vary the light intensity, generally related to the amount of available daylight or occupancy.

Noise attenuation Noise reduction.

Passive building A building which by its configuration eliminates the need for mechanical ventilation, and reduces the need for daytime electric lighting

Photocell Measures illuminance at any position. When placed externally the photocell allows internal light control systems to react to changes in the weather, an element of 'daylight linking'.

Photovoltaics External panels on the southern exposure of a building designed to convert solar energy into useful electricity, a developing technology.

Scene set The use of more complicated electronic controls using a microprocessor, to permit different room appearances to be available at the touch of a button, with a number of scenes being 'preset'; which can subsequently be changed automatically.

Stack effect The way in which hot air will rise in a chimney.

Wind turbine A 'windmill' designed to generate electricity.

7. ARCHITECTURE

Atrium The courtyard entrance to a roman house, with an opening in the centre through which rainwater was collected. This opening also provided light to the courtyard and surrounding rooms. The word has now taken on the meaning of multistorey spaces which are daylit from overhead glazed roofs.

Barrel vault A continuous structural vault of semi-circular section, used from Roman architecture to the present; nowadays formed of reinforced concrete.

Ceiling coffer A form of concrete roof construction, where, to add strength without increased weight, square holes or 'coffers' are omitted leaving a 'waffle' shape into which services can be placed.

Clerestorey (also clear-storey, and pronounced this way) The upper storey with windows above the side aisle roofs, giving high level daylight particularly in a church.

Conservation The protection of works of art against the deleterious effects of the environment. The control of light levels (particularly ultra violet) is a major component of conservation.

Dimensional coordination The manner in which different building materials are planned to fit together.

Folded plate Ceiling development of shell concrete construction.

Glass brick The development of 'bricks' made from glass in the 1930s allowed architects to design structural 'see-through' walls. The Maison Verre in Paris is a well-known modern example; although not widely used today they remain a useful architects tool.

Lighting gantry A light weight 'bridge' independent of the main structure of a building, providing support and electric power to light fittings.

Roof monitors/laylights The roof construction in which daylight is permitted to enter a space from overhead. In the case of factories they would be designed to control the entry of sunlight.

Roof truss A development of the beam supports to a roof allowing an openwork lattice to accept services.

Scale Scale is a matter of 'proportion,' the larger the scale, the less human the building will appear. It is sometimes difficult to judge the size of a particular building or interior until a person is added to give it 'scale'.

Shell concrete A thin skin of reinforced concrete, formed in a curve to span the roofs of large areas.

Sprinkler system Fire control by means of a system of water pipes which are designed to deluge water to douse a fire, when design temperatures are exceeded.

Undercroft A term in mediaeval architecture depicting the lower level vaulting of a cathedral, above which the main edifice is built.

Index of architects and designers

Index of lighting designers

Index of subjects

Entries in **bold** refer to Lighting Projects